Industrial Saundersfoot

Dust Jacket and Frontispiece:
SS Foamville loading coal at Saundersfoot Harbour, 1936
(from a photograph by C. C. Green)

Industrial Saundersfoot

M. R. C. Price

GOMER PRESS
1982

First Impression - April 1982

ISBN 0 85088 866 2

Printed by J. D. Lewis & Sons Ltd.,
Gomer Press, Llandysul, Dyfed

For David and Catherine

CONTENTS

Foreword

I thought I knew something of the story of this area until I read the MS of *Industrial Saundersfoot.*

My great-grandfather, whose own father had worked on the ill-fated Stepaside canal, had the contract for driving stones on one section of the great new road which was built through Stepaside in the 1830's. Later, he had a gang of men, women and children digging and picking iron ore at the Patches between Amroth and Wisemansbridge. As the records show, he had a contract for delivering it to the Grove ironworks.

When his son, my grandfather, married, it was to a girl whose mother and father had both worked underground as children at the New Hayes Pit at Begelly. Not long ago I read the report of an interview with some free-thinking young lady of this enlightened age, whose ambition was to become Britain's first woman miner. I have news for her.

In later years, when sailing boats were still coming into Saundersfoot harbour for coal, my father's old Ford T type lorry was driven by Billy Williams before he became the last driver of the famous ROSALIND. I can remember, as a small boy, going with him in the lorry to fetch the last of the firebricks from the brickworks at Woodside, which had already closed, but where the Foundry was still working.

Inevitably, Martin Connop Price's book will be evocative of such memories and family lore, not only for me, but for so many others of my generation. And that includes visitors from far afield, as well as local people.

This much, however, can be said for many another publication. But never has there been so thoroughly researched a piece of work on the industry which was born, flourished briefly and died as quickly, in this small corner of Little England Beyond Wales.

Not much remains today by way of evidence as to what once was. To have collected, and recorded in book form, so much of its history is cause for gratitude, not only to the author, but to those who have made it financially possible.

Certainly it is the most fascinating and historically valuable book in my native county I have ever read. To have been consulted by the author in the first place was a pleasure, and to have been asked to write this brief Foreword an honour and a privilege.

Roscoe Howells

Introduction

Today Saundersfoot is a village with a busy harbour and a thriving tourist trade. Set amid the natural beauty of south Pembrokeshire, it is hard to believe that the place was brought into being by coal. Tunnels on the coastal path to Wiseman's Bridge betray the old route of the Saundersfoot Railway, and pockmarked cliffs provide a hint of old workings. Even so, most visitors have little idea of the extent of former industry. Nineteen years ago, on my first visit to Saundersfoot, I was no exception. Before long, though, I became fascinated by the story of industrial Saundersfoot. A study of the Saundersfoot Railway was soon under way, and although a booklet appeared in print its inadequacies gradually became more and more evident. By the mid-1970's it was obvious that the subject merited a much larger study, including the coal and iron industries as well as the railway and harbour.

This book is the result. Further research has unearthed a lot of new information, and I have been very fortunate to have the willing assistance of many knowledgeable people. It is quite possible that more material will emerge in time to come, but in the meantime I must express my grateful thanks to all those who have helped me in this study. To record every name would be to set down a list of unmanageable proportions, but the acknowledgements should include all those with whom I have had correspondence. In addition special thanks are due to Ray Bowen, Roscoe Howells and George Edwards for their expert advice and guidance. Dorothy Evans has shown me great kindness in allowing me to see the Vickerman papers, and Trevor Lewis has been similarly helpful in relation to the records of the Saundersfoot Railway & Harbour Co. Without their aid much of this study could not have been written at all. It must also be said that without the very generous assistance of the Welsh Church Fund, the West Wales Association of the Arts, the St. Issell's Community Council, Trevor Lewis and Lady Merthyr, this book could not have appeared in print. I am immensely grateful for their help and support. Margaret Snook has devoted an immense amount of time and trouble to typing the manuscript, and Bob Lyon has given great care to the preparation of the maps. I am greatly indebted to them both, and thankful for their patience. Finally I must say that I have the blessing of a loving and tolerant wife! Without Judy's support this piece of work could have made little progress.

Martin R. Connop Price

October 1981

Wortley, near Sheffield

Note:
Many of the old industrial sites referred to in this book are on private property. References to such sites should not be construed as authority to visit them: in every case permission should be sought from the owner or occupier.

Acknowledgements

Grateful thanks are due to the following establishments and their staff:

National Library of Wales, Aberystwyth.
Dyfed County Record Office, including the Pembrokeshire Record Office, Haverfordwest.
Dyfed County Library, and in particular the Pembrokeshire County Library, Haverfordwest.
Bristol City Library.
Bristol City Record Office.
Cambridge University Library.
House of Lords Record Office, London.
Public Record Office, Chancery Lane and Kew, London.
Patent Office, London.
National Coal Board, South Wales Area Surveyor's Department, Llanishen, Cardiff.
Sheffield City Library.
South Pembrokeshire District Council.

In addition to those mentioned in the Introduction, many individuals have made a particular contribution to this study and their help has been very much appreciated. They include:
Mrs. M. W. Acors, Mrs. M. Beddoe, M. Brace, A. & T. Cole, Mrs. L. Cordon, Mrs. B. N. Davies, Lady Dunsany, D. Elis-Gruffydd, R. W. Ellison, O. Evans, A. J. Flint, D. Geldard, C. C.Green, W. R. Griffiths, O. Gwyther, M. A. Hardy, S. Harris, D. Howell, I. Howell, W. N. Jackson, F. Jones, R. W. Kidner, Mrs. F. Lawrence, A. Lewis, the late E. T. Lewis, G. S. Lewis, J. Pepler, W. Phillips, R. Proser, S. Prout, the late H. Prout, W. H. Rogers, Mr. & Mrs. J. Stredder, G. Thomas, W. Thomas, E. S. Tonks, Mr. & Mrs. F. Usher, G. Watts, P. White, J. Whitehouse, Mrs. A. Williams, P. Winship.

Photographs

C. R. Vickerman (*D. Evans*)
C. H. R. Vickerman (*W. Griffiths*)
Saundersfoot, 1860 (*H. Prout*)
Shipbuilding at Saundersfoot Harbour, c. 1870 (*R. Howells collection*)
East Anglian coal boats, Saundersfoot Harbour, 1870 (*R. Howells collection*)
Saundersfoot Harbour, 1883 (*R. Howells collection*)
Ballast quay, Saundersfoot Harbour, 1898 (*Mrs. A. Williams*)
A schooner running for Saundersfoot in a gale (*W. H. Rogers*)
"Lady of the Isles" aground at Saundersfoot, 1907 (*Mrs. D. Evans*)
View past Bonvilles Court Coal Co. offices towards Railway Street (*W. H. Rogers*)
Bonvilles Court Coal Co. offices after the fire, October, 1913 (*H. Prout*)
Bonvilles Court Coal Co. office staff, 1921 (*Mrs. O. Scourfield*)
The miners' train approaches Saundersfoot harbour (offices in background) (*Mrs. L. Cordon*)
Miners' train halts in Railway Street (*Mrs. D. Evans*)
Miners' train in Railway Street on its way to Stepaside (*Mrs. D. Evans*)

Ruined Blacksmith's Shop, Crickdam Patch, 1978 (*M. R. C. Price*)
Lloyds (or Rooksnest) Patch, 1978 (*M. R. C. Price*)
Step Cottage and old mineyard, (on right) Wiseman's Bridge, 1979 (*M. R. C. Price*)
Kilgetty Colliery site, 1979 (*M. R. C. Price*)
Ruined Engine House, Lower Level, 1979 (*M. R. C. Price*)
Old Engine House at Hacket, 1979 (*M. R. C. Price*)
Site of Broom Colliery, 1978 (*M. R. C. Price*)
Saundersfoot Harbour entrance, 1978 (*M. R. C. Price*)
Bonvilles Court Pithead, 1978 (*M.R. C. Price*)
Former Bonvilles Court Coal Co. cottages, Stepaside, 1979 (*M. R. C. Price*)
Spoil Tips at Moreton Colliery, 1979 (*M. R. C. Price*)
Cleared Site of Broadmoor (Greenhill) Colliery, 1979 (*M. R. C. Price*)

Bibliography and Sources

The starting point for any study of Pembrokeshire coal mining must be George Edwards' excellent little booklet "The Coal Industry in Pembrokeshire", printed from Field Studies Vol. 1, No. 5 (1963). This work was based on the writer's fascinating Birmingham University thesis on the same subject, completed in 1950. A copy of this thesis is to be found in the Pembrokeshire County Library. In the notes which follow these will be referred to as Edwards (1963) and Edwards (1950).

Saundersfoot and district has received some attention in recent years. In addition to my booklet on the Saundersfoot Railway (Oakwood Press, 1964), T. G. Stickings tackles many facets of local history in his volume "The Story of Saundersfoot" (Five Arches Press, 1970), as does D. R. Morgan in "The Story of Begelly" (1980). Roscoe Howells very interesting "Old Saundersfoot" (Gomer Press, 1977) includes many fascinating old photographs. These books will be referred to in the notes as "Stickings", "Morgan" and "Howells".

Two other volumes consulted require abbreviations. One, now hard to come by, is an official publication: "Geology of the South Wales Coalfield, Part IX" (1914). The other is more recent, namely "The South Wales Coal Industry 1841-1875", by J. H. Morris and L. J. Williams (University of Wales Press, 1958). For the purposes of these notes they will be referred to as "G. S. W. C." and "Morris & Williams".

Local newspapers have been a fruitful source of material, although it has not been practicable to make a comprehensive inspection. In the notes the newspapers most frequently consulted may be described by their initials—for example, H. M. H. T. = Haverfordwest and Milford Haven Telegraph; W. T. = Western Telegraph; P. H. = Pembrokeshire Herald; T. O. = Tenby Observer.

The National Library of Wales holds many valuable records relating to industrial Saundersfoot, most notably in the Picton Castle, Haverfordwest (Williams & Williams) and the Eaton, Evans & Williams collections. These will be referred to by the initials "P. C.", "W. W.", and "E. E. W." respectively. The important but uncatalogued Goggerdan papers will be described simply as "Goggerdan". Certain documents in the Bronwydd and Llandinam collections have also been studied.

The Dyfed County archives include the interesting Lowless and Lowless papers (referred to here by the initials "L. L.") and some Carew estate papers. The early minutes of the Saundersfoot Railway & Harbour Co. (hereinafter "S. R. H.") are also in these archives. The National Coal Board, South Wales Area, is in possession of many interesting mining plans. Those studied will be referred to by the initials "NCB", followed by a microprint number, if appropriate.

Finally the privately owned Vickerman papers have been an invaluable source of information. This varied material will be referred to here by the name "Vickerman", and other details as appropriate. Minutes of the Pembrokeshire Iron & Coal Co. may be abbreviated with the initials "P. I. C.".

C. R. Vickerman

C. H. R. Vickerman

Saundersfoot, 1860

Shipbuilding at Saundersfoot Harbour, c. 1870

East Anglian coal boats, Saundersfoot Harbour, 1870

Saundersfoot Harbour, 1883

Chapter One

THE HISTORY IN OUTLINE

Up to 1800

The story of industrial Saundersfoot begins and ends with coal. The records of the Earls of Pembroke contain a reference to coal mining in the Commote of Coytrath as early as 1324[1], long before the village of Saundersfoot was built. At the time a large forest called Coytrath or Coedrath covered much of the district, which was sometimes described by the name of the parish, St. Issells.

Saundersfoot's coal is high grade anthracite, very nearly pure carbon. The coalfield forms a narrow belt running across Pembrokeshire from Saundersfoot Bay in the east to St. Bride's Bay in the west, the seams being an extension of the coal measures in the counties of Carmarthen and Glamorgan. The first treatise on the Pembrokeshire coalfield, written by George Owen in 1603[2], shows that the remarkable qualities of the coal were appreciated at an early date:

"The cole of Pembrokeshire may be nombred as one of its cheefe commodities, and is so necessarie as without it the countrie would be in great distress; it is called stone coal for the hardness thereof, and is burned in chimnies and grates of iron, and being once kindled giveth a greater heete than light, and delighteth to burne in darke places."

Owen went on to praise its smokelessness, and to emphasize that it was "a most proved food dryer of mault, therein passing wood, fern or strawe". He added that in view of its qualities the coal "was carried out of this countrie to the citie of London . . . by a gentleman of experience, to show how farre the same excelled that of Newcastle . . . and if the passage were not so tediouse there would be great use made of it".

The first shipment of coal from Saundersfoot to London is said to have been loaded onto a boat beached at Swallowtree Bay, but at this time much of the output was being dispatched through the port of Tenby or from beaches at Coppet Hall and Wiseman's

Bridge. In spite of the difficulties and distances, by the second half of the seventeenth century Pembrokeshire anthracite was being exported to France, Holland and elsewhere in Europe. It has been estimated that by 1748 some 30,000 tons were being sent abroad each[3] year. At this period, before the great expansion in mining in South Wales, the coal industry in Pembrokeshire was of real importance. The proximity of the sea to most parts of the coalfield was undoubtedly a factor in its early development.

George Owen mentions coal works at several sites in 1603, including Saundersfoot, Begelly and Jeffreston. The scale of the export trade suggests that by the middle of the eighteenth century there was even more activity in the coalfield. Early workings were shallow and inefficient, but they were abundant in the Saundersfoot district. For that reason it is hardly possible to say where the earliest coal mining took place. There are few records of the industry before the late eighteenth century, but from those which do survive it appears that there was a colliery at Wooden in 1756, at Moreton in 1757 (and probably earlier) and one on the Hean Castle estate near Coppet Hall by 1764. A shaft was sunk at Kilgetty in 1775, and at the end of the following year quite extensive mining was under way at Begelly and Ridgeway. In 1784 a colliery was opened at Thomas Chapel and in 1786 there were active pits at Stonecross, near St. Issells church. In the 1780's two shafts at least reached a depth of 84 yards, although generally speaking deeper mining had to await the introduction of new technology early in the nineteenth century.

Until the first half of the nineteenth century coal mining was closely associated with agriculture. Charles Hassall, writing in 1794, took a dim view of this relationship[4]. He declared that the collieries "have a material influence in checking the progress of agriculture in the part of the county through which they pass. No canal navigations, no railways have yet been established for carrying the coal and culm of these collieries to the shores and creeks where they are shipped. All the carriage is done by carts, drawn by two oxen and two horses to the total ruin of the roads". Coal mining undoubtedly had an adverse effect on agriculture. According to the Crop Returns of 1801 the amount of corn sown in the parish of St. Issells was reduced by the need to use land

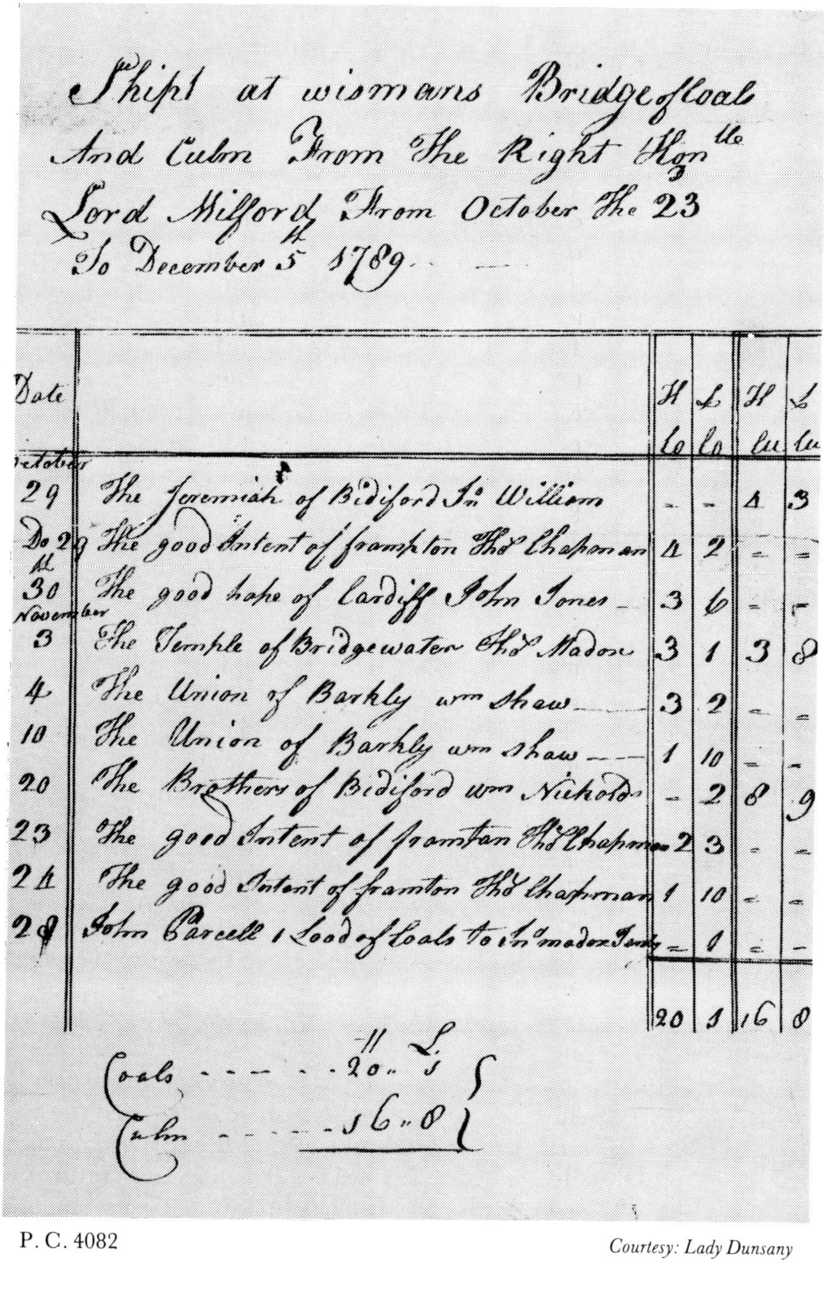

Shipt at wismans Bridge of Coal And Culm From The Right Honble Lord Milford From October The 23 To December 5 1789 . . —

Date		H Co	L Co	H Cu	L Cu
October 29	The Jeremiah of Bidiford Jn William	–	–	4	3
Do 29	The good Intent of frampton Thos Chapman	4	2	=	=
Do 30	The good hope of Cardiff John Jones —	3	6	=	– [?]
November 3	The Temple of Bridgewater Thos Madox	3	1	3	8
4	The Union of Barkly Wm Shaw —	3	2	–	–
10	The Union of Barkly Wm Shaw —	1	10	–	–
20	The Brothers of Bidiford Wm Nicholds	–	2	8	9
23	The good Intent of framtan Thos Chapman	2	3	–	–
24	The good Intent of framton Thos Chapman	1	10	–	–
28 [?]	John Parcell 1 Lood of Coals to Jn madox Tenby [?]	–	1	–	–
		20	1	16	8

	H	L
Coals	20	1
Culm	16	8

P. C. 4082

Courtesy: Lady Dunsany

"for grazing the teams of cattle that convey the coal and culm to the shore". In the parish of Begelly the situation seems to have been worse, because the corn raised was said to be "not nearly sufficient for the consumption of its inhabitants". The explanation provided was interesting. "The parish lies in the midst of coal works, and the landholders, whose farms in general are small, are chiefly engaged in leading coal and culm to the shore, which they are bound to do by a covenant in their leases. Depending principally upon this business for their livelihood they greatly neglect the cultivation of their farms. Many keep a team of two horses and two oxen, and some keep two teams, who do not sow a sufficient quantity of corn for their own families."[5]

It is clear that at the end of the eighteenth century industry in Pembrokeshire remained relatively primitive. The main problem was the inability or unwillingness of landowners to put large amounts of capital into coal mining. Many of the pits were small concerns opened up on lease by a few workmen, and they had a constant struggle to pay their rents and royalties at all, let alone step up their investment. Some landowners became more involved, but usually only for their own benefit. In the Saundersfoot area, for example, John Wogan was working coal on the Hean Castle estate, and Edward Loveden was working pits on Lewis Pryse's property at Moreton. The Philipps family of Picton Castle were involved in mining on their lands at Begelly and Kilgetty, having earlier given up a colliery at Moreton following one of the not unusual arguments between rival proprietors.

Transport Improvements

As Charles Hassall suggested, improvements in transport were essential. In 1794 he was able to report that Sir Richard Philipps, Lord Milford, had made an unsuccessful attempt at canal building between the collieries at Kilgetty and the coast at Wiseman's Bridge. The Kilgetty canal is believed to have been the only canal project ever undertaken in Pembrokeshire.

The first proposals for the building of tramroads or railways made no progress. In 1797 and 1806 the mining engineer John Curr advocated the construction of tramroad between Moreton and Saundersfoot, a distance of about one mile. It seems that

Lord Milford was obstructive, and nothing was done. In 1824 a scheme for a tramroad between Begelly and Saundersfoot also failed, but at this period there is evidence of a tramroad between St. Isells and Coppet Hall, along the route now known as the Black Walk, all of which was then within the Hean Castle estate. Unfortunately no dates for the construction and opening of this line have come to light.

Following the death of Lord Milford in 1823, Sir Richard Bulkeley Philipps succeeded to the Picton estate, and soon showed an interest in its development by approving the 1824 tramroad scheme. At the same time he may have become involved in the running of collieries on the estate at Kilgetty. By 1828 Sir Richard Philipps and his solicitors, the trustees of Lord Milford's estate, were working on a plan for a new railway and harbour at Saundersfoot. They presented their project to Parliament, and in 1829 an Act was passed incorporating the Saundersfoot Railway & Harbour Co[6]. This Act envisaged the construction of over four miles of railway between collieries at Thomas Chapel and a substantial stone built harbour at Saundersfoot. Two branch lines were also authorised, one between Saundersfoot and Merton (or Moreton) and one running north along the coast to Wiseman's Bridge. It was an ambitious project, and the willingness of Sir Richard and his colleagues to invest heavily in the company as well as in the Picton estate provided an important impetus for growth. Indeed the S. R. & H. Co. was really the first scheme to benefit the whole district rather than just one property.

Building the Harbour and Railway

The S. R. & H. Co. soon discovered that building the harbour was no easy matter. At the time of the Act only one outer pier was deemed necessary, but before long an inner pier was proposed to counteract the effects of sea swell. Storms were a constant hazard. Every winter the works were battered by heavy seas; every spring R. W. Jones, the engineer, reported the need for repairs. Fortunately construction of the railway was simpler, and by 1833 the first short sections of track had been laid to help convey rocks to the harbour site. The principal features of the Thomas Chapel line were a rope-worked incline a short distance inland from

Saundersfoot, and a tunnel further west to carry the railway through to Kingsmoor. On the Wiseman's Bridge branch there were three tunnels through the cliffs in the vicinity of Coppet Hall. No work was done on the branch line to Moreton.

The harbour and horse-worked four foot gauge railway were ready for use in 1834, although the Wiseman's Bridge branch was beset by storm damage, and failed to work regularly until 1836. It was then extended to the Lower Level colliery near Kilgetty. Initially the company was successful, and business grew steadily during the 1830's. By the end of the decade plans were afoot for the improvement of both harbour and railway. For this purpose the company obtained two Acts of Parliament in 1842. The Act dealing with the harbour contemplated the construction of a new wet dock[7]. The Act dealing with the railway included powers for the extension to Lower Level, the branch to Moreton, and another branch to Broadmoor and Masterlands[8]. Apart from the Lower Level extension, already largely complete, nothing came of these new proposals.

Local industry was by no means confined to the company and the Picton estate. The long established collieries on the Hean Castle estate and at Moreton continued to flourish. The Child family worked pits at Begelly, and there were other pits at Broadmoor and Jeffreston in the west and at Amroth in the east. Indeed as records are not plentiful it is easy to underestimate the extent of coal mining at the period. Even in 1794 Hassall was able to record that 120,000 tons of coal were shipped from Pembrokeshire in a year, all but 12,000 tons being British coastal traffic. Allowing for the substantial quantities used in the county for limeburning, smithy work and domestic purposes, the total output for the year was probably well over 150,000 tons. Saundersfoot's share in the trade cannot be assessed accurately, but it seems to have jumped following the completion of Saundersfoot Harbour. In 1833 11,497 tons of coal and culm were exported[9] from Saundersfoot, but by 1840 the total was 39,405 tons. Relatively few figures are available but thereafter it appears that for 35 years or so annual shipments averaged 30,000—40,000 tons.

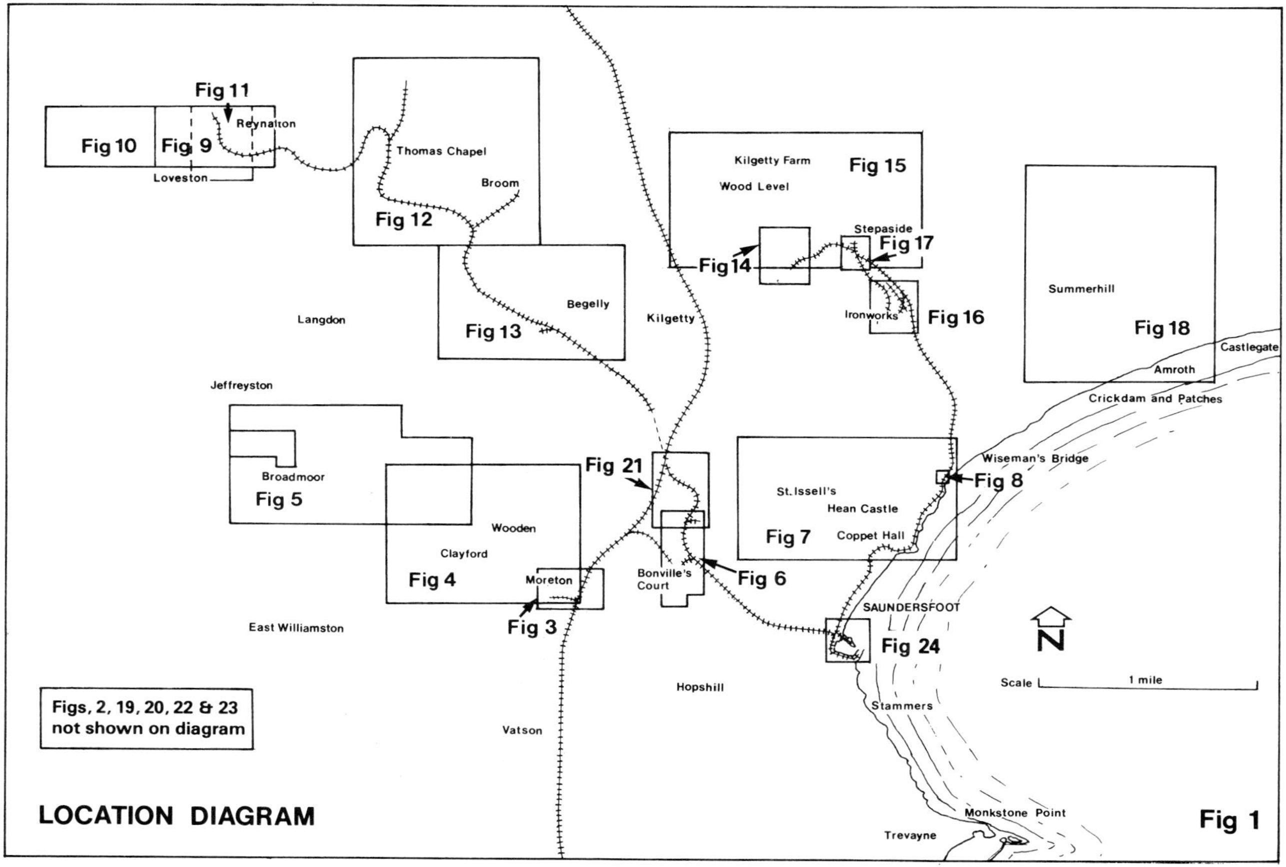
LOCATION DIAGRAM
Fig 1
Figs, 2, 19, 20, 22 & 23
not shown on diagram
Fig 10
Fig 9
Fig 11
Reynalton
Loveston
Thomas Chapel
Broom
Fig 12
Begelly
Fig 13
Langdon
Kilgetty
Kilgetty Farm
Wood Level
Fig 15
Stepaside
Fig 17
Fig 14
Ironworks
Fig 16
Summerhill
Fig 18
Castlegate
Amroth
Crickdam and Patches
Jeffreyston
Broadmoor
Fig 5
Wooden
Clayford
Fig 4
Moreton
Fig 3
Fig 21
Bonville's
Court
Fig 6
St. Issell's
Hean Castle
Coppet Hall
Fig 7
Wiseman's Bridge
Fig 8
SAUNDERSFOOT
Fig 24
N
Scale
1 mile
East Williamston
Hopshill
Stammers
Vatson
Monkstone Point
Trevayne

The Heyday of Saundersfoot's industry

By now steam engines for pumping and winding were widely available, and it was becoming practicable to sink shafts to deeper coal seams. In the 1840's a number of shafts were sunk, and most notably two at the new Bonvilles Court colliery. Under the direction of J. P. M. Myers this pit soon won an enviable reputation as a source of top quality anthracite. Moreton colliery meanwhile was developing under Messrs. Williamsons, the new lessees, and on the Hean Castle estate Thomas Stokes was working both coal and iron ore. In about 1845 he also founded a fire-brick works at Woodside, near Wiseman's Bridge. A little earlier, between 1840 and 1842 the collieries at Begelly, Broadmoor, Hean Castle, Kilgetty and Thomas Chapel were visited by R. H. Franks, on behalf of the famous Children's Employment Commission[10]. To judge from his report, conditions were poor, even by the dismal standards of the time. Not only were small children employed underground, but women in the mines sometimes had to work "harder than the slaves in the West Indies". Poverty and ignorance were commonplace.

The collieries of the Picton estate did not escape Franks' comments and criticisms, but Sir Richard Phillips and his lawyers continued to provide most of the drive for new development. In 1845 they asked Messrs. Fosters to study the minerals of the Saundersfoot district and make recommendations on the exploitation of the reserves of iron ore and anthracite. After a careful survey Messrs. Fosters advocated the construction of an iron-works with four blast furnaces using local coal[11]. They suggested a site for the works at Saundersfoot, and having described the anthracite of the Kilgetty vein as "first rate—the purest anthracite in South Wales", recommended that two new collieries be sunk to work it, one at Kilgetty and one at Thomas Chapel.

This report was a landmark in the history of Saundersfoot's industry, because it drew attention to the local potential for iron-making. Hitherto any ore that had been mined had been dispatched to ironworks on the other side of Carmarthen Bay. Now it was recognised that the iron (or "mine", as it was usually called) might be used in a local ironworks. After attending to some preliminaries, Sir Richard and his advisers set up the Pembrokeshire Iron & Coal Co. in June, 1847. Although they obviously had Fosters' report in mind they did not follow the recommendations exactly. The most important difference was that they decided to establish their ironworks near Kilgetty.

At the same period Sir Richard and his colleagues helped to promote the Tenby, Saundersfoot & South Wales Railway & Pier Co. which was incorporated by Act of Parliament in 1846[12]. This was intended to construct a line from Tenby and Saundersfoot to join the Pembroke Dock branch of the recently authorised South Wales Railway at Reynalton. For this purpose the T. S. & S. W. R. & P. Co. proposed to take over the S. R. &. H. Co. and rebuild much of the railway to the S. W. R.'s 7ft 0¼in. gauge. It was confidently predicted that if the plans were put into effect "every spot of land" in Saundersfoot would soon be covered with houses. In the event nothing happened. The S. W. R.'s Pembroke Dock branch was not built, and without it the proposals of the T. S. & S. W. R. & P. Co. were pointless. By June 1847 its purchase of the S. R. & H. Co. had been postponed, never to be revived. In spite of this setback the village of Saundersfoot continued to grow.

The Pembrokeshire Iron & Coal Co.

The construction of Kilgetty ironworks began soon after the launching of the P. I. & C. Co. By 1849 two blast furnaces were ready for operation, and one was put into blast. Iron ore was obtained from the cliffs and from several newly opened levels; coal came from the pits at Kilgetty and Lower Level. Unfortunately allegations about the misuse of funds led to a terrific boardroom row, litigation ensued, and several prominent directors departed. The company was left in the care of the same group of Picton estate advisers who already controlled the S. R. & H. Co. The most notable of these was C. R. Vickerman.

An explosion at the ironworks and a slump in trade brought ironmaking to an abrupt halt early in 1850. Plans to build more furnaces were abandoned. In 1852/3 there was another attempt at ironmaking, but it was no more successful. In the meantime the company produced coal, and in 1852 acquired collieries at Thomas Chapel. Between 1853 and 1859, in anticipation of better days, the Grove colliery was sunk on the hillside behind the ironworks to enable it to feed coal directly into the furnaces. To the consternation of all concerned Grove proved to be costly in terms of life and capital. A new colliery at New Hayes, near Thomas Chapel, was similarly expensive, and less productive. Even so the company resumed iron making at Kilgetty in 1857 and struggled on until 1861.

By 1861 the P. I. & C. Co. was controlled by one man, Charles Ranken Vickerman. As losses mounted he persevered whilst his colleagues dropped out, and that in spite of the fact that the Rev. J. H. A. Philipps, who had succeeded to the Picton estate in 1857, was less sympathetic than his predecessor Sir Richard Philipps (later Lord Milford)[13]. In truth C. R. Vickerman was building a small empire; by 1863 he had gained control of Bonvilles Court colliery and the Saundersfoot Railway & Harbour Co. In September, 1863, he purchased the Hean Castle estate, and became the pre-eminent figure in Saundersfoot[14].

Decline sets in

The involvement of C. R. Vickerman gave rise to further investment and a degree of industrial activity which was not entirely typical of anthracite producing areas at the time. By 1860 most ironworks in such areas were in a depressed condition, and reckoned to be poorly placed for raw materials and good communications. However in 1863, after an interruption of two years, Kilgetty ironworks resumed production. Some pig iron was made for local use, for example at the Woodside Foundry at Wiseman's Bridge, but most was shipped away. In 1864 almost 3,000 tons were dispatched from Saundersfoot Harbour. Total output was still small in comparison with a great ironworks like Dowlais, and the pressure of competition was considerable. When production stopped for repairs in 1867 it was not restarted on any scale for several years. The small army of men and women who toiled to dig iron mine on coastal cliff patches, or in levels, saw most of their output go for export. Before 1870 the iron industry, which had never been strong, was in decline.

In spite of problems in the iron trade, the outlook for anthracite mining may have seemed more encouraging. Pembrokeshire anthracite was held in high esteem: its suitability for drying malt was well established, and various naval authorities recognised the value of the coal as a clean and economical fuel. Brunel is believed to have designed the steamship "Great Eastern" with this anthracite in mind; Queen Victoria is said to have been so pleased with smokefree Saundersfoot coal that she ordered that no other be burned in the Royal Yacht when she was on board[15]. Unfortunately, when properly alight, anthracite generated too great a heat for ordinary firebars and boilers of the period. Other grades of Welsh coal were soon seen to be better suited for raising steam. On the Continent, however, and especially in Scandinavia, anthracite became popular for use in slow-burning domestic stoves. In Britain it was used as a domestic fuel predominantly in South and West Wales, close to where it was found. Otherwise anthracite was restricted largely to the brewing and lime burning markets where other coals had proved unsatisfactory. Demand for these purposes expanded very little.

The overall picture was now one of no growth. Indeed the available evidence indicates that total anthracite production in 1864 (probably just over half a million tons) was actually lower than the output ten years earlier[16]. In 1865 Pembrokeshire produced 146,000 tons of coal, representing 29% of South Wales anthracite output[17]. Saundersfoot and Milford Haven shipped 82,000 tons of this total, an amount significantly more than was then shipped through Swansea. Thereafter Pembrokeshire's share of the trade declined steadily. By 1874 the volume of shipments from Milford had dropped to less than half those of thirty years before[18]. Saundersfoot's share fell too, although maybe more gradually. The writer Mary Curtis actually mentions an increase in trade at Saundersfoot, at least in the 1870's[19], but the overall picture was one of decline. Rising output in the counties of Carmarthen and Glamorgan intensified competition at a time when good, workable coal was becoming increasingly hard to find in Pembrokeshire. Although the opening of the Pembroke and Tenby Railway through the Saundersfoot district in 1866 improved communications, it also made local markets more vulnerable to imported coal. The outstanding qualities of Pembrokeshire anthracite became the main buttress against competition.

For some collieries this was not enough, and in this period pits at Begelly and Amroth and elsewhere ceased work. At Bonvilles Court and Moreton, however, there was sufficient optimism to enable new equipment to be installed and new workings developed.

The Bonvilles Court Coal & Iron Co. Ltd.

Notwithstanding investment by C. R. Vickerman, the ironworks and almost all the collieries suffered from a shortage of capital. Coal mining in Pembrokeshire was never simple: narrow seams, geological faults and poor drainage were recurring difficulties giving rise to increasingly high costs. Ironmaking, like coal mining, was affected by Pembrokeshire's comparative isolation, away from the main centres of South Wales industry. Furthermore the ironworks came into being at a time of considerable technological change, and advantages perceived in 1845 were much less obvious by 1875. Although there was no lack of expert opinion on what might be done at Kilgetty, in retrospect it is clear that the innovation and re-equipment carried out was too little and too late.

The fact that one man owned the ironworks and most of the collieries was no great advantage when the individual concerned had already lost money. Over the years C. R. Vickerman had sunk huge sums into his business and seen remarkably little return. As time went by he became increasingly aware that his pocket was not bottomless. In 1872, apparently weary with the anxieties and responsibilities of it all, he decided to sell his industrial interests in the expectation that a new company would be formed to take the burden of the business and provide fresh capital. Freed from such worries, there must be more than a suspicion that C. R. Vickerman hoped to establish himself as a landlord receiving royalties and rents rather than a coal-owner carrying a great deal of risk. In 1872 Vickerman purchased the Moreton and Trevayne estates near Saundersfoot and soon after made plans for the reconstruction of Hean Castle.

James Carlton of Knutsford, Cheshire, agreed to purchase the Railway & Harbour company, Bonvilles Court colliery and the Kilgetty collieries and ironworks in June, 1872. After some haggling over the price, Vickerman received £144,140 and retained an interest in the business. Carlton's purchase paved the way for the establishment of the Bonvilles Court Coal & Iron Co. in May, 1873, and some much needed new investment. Part of the Saundersfoot Railway was rebuilt as a locomotive line, and at the ironworks a furnace was put back into blast. Unfortunately this had hardly been done before the new company encountered all the problems of an economic depression. By 1874 it had deepened, and soon the miners and labourers around Saundersfoot suffered wage cuts and acute poverty. In June, 1874, the ironworks' furnace was blown out, ending (as it transpired) the production of pig iron at Kilgetty. Prices continued to fall and at the end of 1876 the B. C. C. & I. Co., beset by labour troubles, could take no more. The company went into liquidation, although the liquidator under sanction of the Court of Chancery, continued to run the collieries only.

The property and plant of the B. B. C. & I. Co. was offered for sale at an aution held in Tenby on 17 May, 1878. The only bidder was the company's largest creditor, C. R. Vickerman, who was owed £100,000 and more. In view of his involvement the liquidator and the court accepted an offer of just £10,000 for the lot[20]. Accordingly C. R. Vickerman achieved the remarkable feat of regaining possession of the works just as if he had never sold them, but reaping the benefit of all the capital expenditure made since 1872—a matter of £40,000 or £50,000.

The Last Years of the Nineteenth Century

By 1878 industrial Saundersfoot had lost not only the ironworks, but also most of the collieries. Vickerman had three pits in a workable state, Bonvilles Court, Grove and Lower Level, of which only Bonvilles Court produced much anthracite. In spite of the depression Messrs. Williamsons kept Moreton in production, and there was a small output from Shipping colliery near Thomas Chapel, where Joseph Cadman was in charge. Both the Williamsons and Cadman now held leases from C. R. Vickerman, who once again totally dominated Saundersfoot's industry. For all his monopolistic power, however, Vickerman was not without worries. By March, 1879 his properties were mortgaged to the hilt. A land agent who met him at the time observed that he seemed "quite broken hearted".[21]. Losses with the B. C. C. & I. Co. were only partly to blame. The reconstruction of Hean Castle and the costs of some litigation had helped to create debts of £90,000. Not surprisingly, the land agent advised C. E. G. Philipps, the latest inheritor of the Picton estate, not to press Vickerman too hard for royalties and rents.

Vickerman's solution to his problems was to try to sell his properties once again. In this he was singularly unlucky. In 1878 negotiations were going well with Mr. Gretton of Bass & Co., one of his anthracite customers, when Gretton's expert mineral surveyor died suddenly. Gretton abandoned his purchase, and left Vickerman looking for a buyer[22]. He was still looking two or three years later, in spite of a few enquiries. Eventually a sale seemed so unlikely that he decided to plan for a future with his not very profitable possessions. Any hopes he had for reviving the ironworks now faded, but he sought reports on local mineral reserves and obtained a number of new coal mining leases. In 1882-1884 he supervised the sinking of Foxholes Colliery, near Bonvilles Court, but once complete its career was regrettably brief. The general economic outlook remained uncertain, and a large part of the local population still suffered considerable poverty. Industrial decline continued on its cheerless way. Shipping colliery closed in 1881 and Moreton colliery in 1887; it is thought that Grove colliery ceased to wind coal in 1884, although its workings were linked with Lower Level colliery by a fine underground roadway built in 1878/79.

In the 1880's C. R. Vickerman endured one difficulty after another. The economic depression and increased competition were bad enough, but setbacks in litigation and some loss of trusted staff were additional blows[23]. Worst of all his wife died in February, 1885, and for the first time he considered selling Hean Castle. Particulars for sale were prepared at the price of 150,000 guineas, but no sale took place[24]. Instead, by 1886 C. R. Vickerman had had the Hean Castle, Moreton and Trevayne estates valued as security for further mortgages, and he had decided to consolidate what was left of his empire. Accordingly in or about 1888 much of Kilgetty ironworks was dismantled and the plant sold. Wood Level, near Lower Level, was re-opened and worked for a number of years, and in 1890-1891 new mineral leases were obtained in an effort to make the most of reserves. At the same period the railway between the harbour and Bonvilles Court incline was relaid for locomotive use; by then the railway west of Saundersfoot station had been lifted.

C. R. Vickerman died in January, 1897. In years he was almost the exact contemporary of Queen Victoria, but although his death preceded hers by four years, in the Saundersfoot district his passing (like hers) brought an era to an end[25]. His son Charles Henry Ranken Vickerman inherited the collieries and commercial interests, and probably most of his father's debts. In any event, in 1899 he sold Hean Castle and its grounds to Sir William Thomas Lewis, the first Baron Merthyr of Senghenydd, and a prosperous coal owner in Glamorgan. The Vickerman family moved to St. Issell's House near the harbour in Saundersfoot. If this was evidence of the family's changing fortunes, the closure of Lower Level colliery soon after was evidence of the unhappy state of local anthracite mining. In 1890 Pembrokeshire produced 71,908 tons, or 6% of South Wales anthracite output; by 1920, 53,240 tons were produced representing a mere 1.2%[26]. Even so, at the turn of the century the industry in Saundersfoot was far from finished.

After 1900

In 1900 Bonvilles Court was easily the largest colliery in the county, and the only pit of any size in the Saundersfoot district. Following the closure of Lower Level an attempt was made to re-open the old Kilgetty colliery, but although men were employed there for several years the scheme could not be called a success. Most of those made redundant at Lower Level went to work at Bonvilles Court. Thereafter, apart from the miners' strike of 1912 and a serious fire at the company's offices in 1913, the years up to World War 1 were relatively uneventful.

By 1914 plans were afoot for renewed mining at Reynalton, west of Thomas Chapel. The New Reynalton Anthracite Coal Co. was formed, and in 1915 the Saundersfoot Railway was relaid and extended to serve the new colliery. A locomotive was provided, and apart from an interruption in 1918 coal mining continued for several years. In 1921, however, labour troubles brought the colliery to a standstill, and subsequently it was abandoned. Meanwhile at Broadmoor and elsewhere in the Saundersfoot district small groups of colliers continued the custom of crop-working, or extracting coal from shallow pits. This was particularly popular in and after 1926 when Bonvilles Court was closed for several months by a miners strike. Already in 1924 the last small remnant of the iron industry had gone with the closure of the Woodside Foundry. Now the prevailing economic gloom darkened the prospects even for Bonvilles Court. By 1929 the company, under the direction of Capt. J. F. Vickerman, was struggling to survive. Wages cuts were called for, but plainly the pit was unprofitable. In April, 1930, Bonvilles Court colliery shut down, putting three hundred men out of work. It was a body blow for Saundersfoot and district.

In spite of this catastrophe the Bonvilles Court Coal Co. was not wound up, and J. F. Vickerman and his associates anxiously considered ways of putting the company back on its feet. In 1932 it was decided to redevelop Broom colliery, between Begelly and Thomas Chapel, and in the following spring work began on pumping out the pit. Production started in 1934, and in the same year plans to resuscitate Kilgetty colliery were put in hand. Coal mining at Kilgetty resumed late in 1936, by which time Broom colliery was approaching an output of 30,000 tons per annum[27]—a figure comparable to production at Bonvilles Court in its last years. Unfortunately 1936 was marred by a disaster at the small Loveston colliery, which had been re-opened in 1932. An inrush of water from old workings cost the lives of seven men, and provided a sharp reminder of how extensively the district had been worked in earlier days.

The coal industry revival was sadly short lived. It reached its zenith in 1938 when Broom produced over 25,000 tons and Kilgetty 16,000 tons[28]. By the end of the year, however, the threat of closure was very real. Although there was talk of a reprieve, both pits shut down early in 1939, and the Saundersfoot Railway promptly followed. This was almost the close of the story, but in 1944 a group of colliers re-opened Wood Level again. It was

Ballast quay, Saundersfoot Harbour, 1898

A schooner running for Saundersfoot in a gale

''Lady of the Isles'' aground at Saundersfoot, 1907

View past Bonvilles Court Coal Co. offices towards Railway Street

Bonvilles Court Coal Co. offices after the fire, October, 1913

Bonvilles Court Coal Co. office staff, 1921

The miners' train approaches Saundersfoot harbour (offices in background)

Miners' train halts in Railway Street

Miners' train in Railway Street on its way to Stepaside

worked as a small private coal mine for several years, but in 1950 it was abandoned bringing coal mining in the district to an end[29]. Industrial Saundersfoot was no more.

Footnotes

[1]Accounts of the Earldom of Pembroke, quoted by H. Owen (1918) A Calendar of Public Records relating to Pembrokeshire: Pembroke and Tenby. Cymmrodorion Record Series, VII, iii, 84, 124, 130, 140.

[2]George Owen (1603) "The Description of Pembrokeshire" Cymmrodorion Record Series, I.

"The Description of Pembrokeshire" Cymmrodorion Record Series, I.

[3]Quoted by Stickings, p. 23. Much of this total was shipped from Cresswell Quay, Landshipping and other small quays on the River Cleddau.

[4]Charles Hassall (1794) "General View of the Agriculture of the County of Pembroke", p. 59, London.

[5]"The Acreage Returns of 1801 for Wales"—D. Williams, Bulletin of the Board of Celtic Studies, Vol. XIV parts i-ii (1950 and 1951).

[6]10 George IV Cap. 108/1829.

[7]5 Victoria Sess. 2, Cap. 53/1842.

[8]5 Victoria Sess. 2, Cap. 35/1842.

[9]H. L. V. Fletcher (1969) "The Coasts of Wales", p. 74. In 1835 the Tenby & Begelly Coal Co. raised 16,232 tons and shipped 12,040 tons through Saundersfoot (Edwards, 1950).

[10]The Royal Commission inquiring into the employment and condition of children in mines and manufacturers was established in October 1840, and reported in 1842. R. H. Franks was one of the inspectors for the Commission in South Wales.

[11]Fosters' Report : PC 4099.

[12]9 and 10 Victoria Cap. 256/1846.

[13]Lord Milford's successor, the Rev. J. H. A. Gwyther, of Madeley, Shropshire was given licence to take the name Phillipps. Relations with the Picton estate advisers may have cooled as a result of an 1859 argument, apparently over the late Lord Milford's shares in the ill-fated T. S. & S. W. R. & P. Co.

[14]Hean Castle was sold by auction at the Gate House Hotel, Tenby, on 9 September, 1863 (Advert in HMHT, 15.7.1863).

[15]Letter of C. R. Vickerman to C. H. Perkins, Sketty, Swansea, dated 5.8.1880.

[16]Evidence of Thomas Evans, Royal Commission on Coal Supplies, 1871, Vol. 2, 18.5.1867, quoted by Morris & Williams, p. 46.

[17]R. Hunt (1865) Mineral Statistics of the United Kingdom, London.

[18]Morris & Williams, p. 93.

[19]Mary Curtis (1880) "Antiquities of Laugharne, Pendine and their neighbours", p. 331/2. "Fifty years ago there was a considerable trade carried on, which fell off. Seven or eight large ships would be generally anchored in the bay. Now trade is returning to it, and the buildings have considerably increased in the last eight years. In the summer the visitors are numerous. There is a harbour and hotels." Even in 1880 tourism was established in Saundersfoot!

[20]Vickerman: memo showing position of works and ownership, 9.1.1882.

[21]PC 3887/8; Letter of S. H. Owen to C. E. G. Phillips, 15.3.1879.

[22]Vickerman: W. S. Gresley's Report, 9.11.1878.

[23]The death of Rev. J. H. A. Phillipps in 1876 led to a tremendous argument over the succession to the Picton estate. For C. R. Vickerman the failure of the Bonvilles Court Coal & Iron Co. was more serious, not only because his own investment was at risk but because W. T. Longbourne withdrew his money in such a way as to leave him liable on some very large debts. C. R. Vickerman appears to have been either unlucky or unwise in the choice of some of his key personnel. William Foley, who enjoyed advancement with Mr. Vickerman for almost thirty years and eventually became works manager for the Bonvilles Court Coal & Iron Co., left Saundersfoot in 1878. Subsequently it was alleged that he had had his fingers in the till for some time. In 1881 the clerk to the Bonvilles Court company, F. Barnaschone, admitted embezzlement, and in 1885 O. Adams was discovered committing a similar offence. Fortunately David John, a trusted assistant in the 1870's and later manager of the Bonvilles Court, Grove and Kilgetty collieries, managed to leave with his reputation intact. Before his departure to become general manager of collieries near Bridgend the workmen gave him a gold watch at a dinner in his honour (T. O., 5.12.1882).

[24]Vickerman: particulars of Hean Castle estate, 11.7.1885.

[25]The funeral took place at St. Issells on 26.1.1897, and it was estimated that a thousand people were present in or outside the church. Although a press report referred to "the general sorrow that prevailed" it cannot be said that C. R. Vickerman was a very popular employer. He was uncompromising with those who let him down, and firm in upholding his own rights. Equipped with a lawyer's grasp of language and argument, he was persistent in pressing his own case and eloquent in the rebuttal of others. This and the steady growth of his own power in the district, was never likely to endear him to all. The fact that his immense investment largely sustained the local economy was often overlooked.

Criticism of C. R. Vickerman as a landlord rests primarily on evidence given to the Royal Commission on Land in Wales at its hearings in Pembrokeshire in 1894. Mr. Vickerman dealt with some damaging allegations but felt that he did not get a fair hearing. There is some reason to believe that the Royal Commission was less impartial than such august bodies are supposed to be: H. M. Vaughan (1926) "The South Wales Squires" argues that it was established for political reasons and given a membership largely unsympathetic to landlords. Arguably C. R. Vickerman was no better and no worse than any other landlord of his time.

In spite of all the setbacks, C. R. Vickerman's little empire at Saundersfoot provided him with the setting for the life he liked to lead—unlike the irksome routine of a solicitor's office. It was a life with the freedom to be his own master, with ample opportunity for entertainment. Even in old age C. R. Vickerman was an indefatigable sportsman, and yet he was still energetically involved in colliery business. His papers do not suggest that his disappointments ever lowered his zest for life!

[26]Edwards (1963) p. 28.

[27]Edwards (1963) p. 14.

[28]Edwards (1963) p. 19.

[29]Hook Colliery, West Drift, south-east of Haverfordwest, was the only Pembrokeshire pit to be worked by the N. C. B., but it was closed in April, 1948. In 1950 work began on a new drift at East Hook, near the River Cleddau, but the venture appears to have been abandoned before any coal was mined.

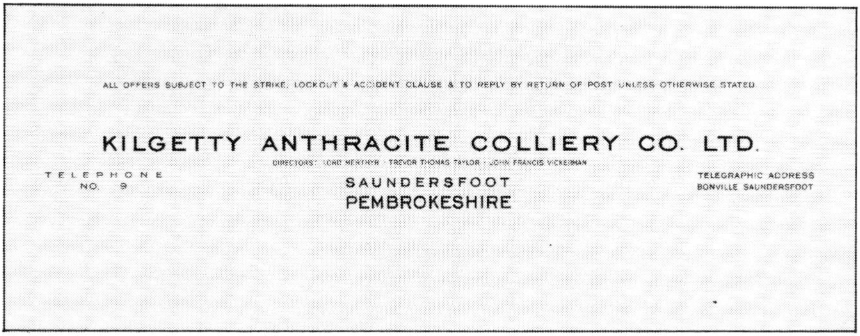

ALL OFFERS SUBJECT TO THE STRIKE, LOCKOUT & ACCIDENT CLAUSE & TO REPLY BY RETURN OF POST UNLESS OTHERWISE STATED

KILGETTY ANTHRACITE COLLIERY CO. LTD.

DIRECTORS: LORD MERTHYR · TREVOR THOMAS TAYLOR · JOHN FRANCIS VICKERMAN

TELEPHONE
NO. 9

SAUNDERSFOOT
PEMBROKESHIRE

TELEGRAPHIC ADDRESS
BONVILLE SAUNDERSFOOT

635 Gt. Western Ry. Gt. Western Ry 635
Whitland Whitland
TO
SAUNDERSFOOT
THIRD CLASS
2/1 P Fare 2/1 P
Saundersfoot Saundersfoot
FOR CONDITIONS SEE BACK E.B

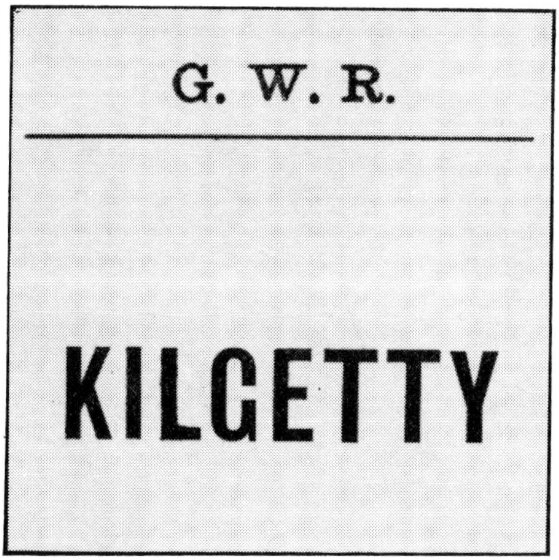

G. W. R.

KILGETTY

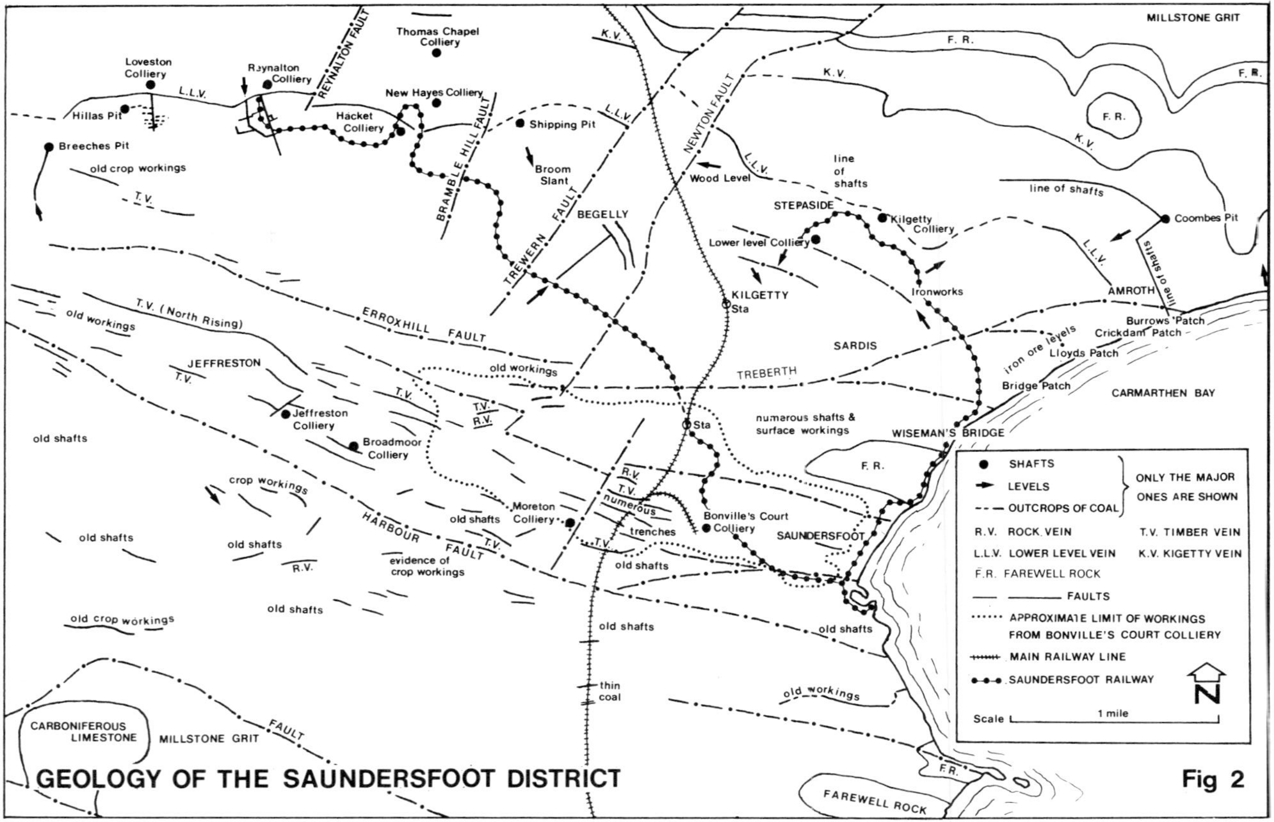

GEOLOGY OF THE SAUNDERSFOOT DISTRICT

Fig 2

Chapter Two

THE COAL INDUSTRY:
The Technical and Social Background

Geology

The coal measures of Pembrokeshire lie across the county in the general form of a syncline or trough, with millstone grit and carboniferous limestone outcropping both to the north and the south. This synclinal form is due to the coalfield having been involved in severe geological disturbances in the Armorican period, after the carboniferous rocks were laid down. Considerable folding and faulting occurred, creating a complex and irregular geological structure in which repeated overthrusts from the south had a particularly marked effect, distorting the strata and sometimes pushing it towards the vertical. As a result the higher seams of coal were frequently shattered, and reduced to culm (small crumbly fragments of coal). The lower seams were much less damaged, although the main faults were clearly discernible at depth.

In the Saundersfoot district the syncline in the coal measures can be found, but the prevailing dip of the strata is to the south. This is interrupted by two major faults, both of which appear as downthrows to the north. The first, named the Erroxhill Fault, runs from the coast near Amroth due west to Sardis Mountain and along the south side of Kingsmoor. The throw of the fault is estimated at 93 yards near Sardis Mountain, but at more than 200 yards beyond Treberth, near the junction with the much smaller Wiseman's Bridge fault. The second fault is called the Harbour Fault, and this runs from the vicinity of Saundersfoot Harbour east north-east past Moreton and Broadmoor. Together these faults have the effect of dividing the Saundersfoot district into three distinct geological areas—one north of the Erroxhill fault, another between the Erroxhill and Harbour faults, and the third south of the Harbour fault. However as the faults downthrow to the north, and the seams or veins of coal rise in that direction, the veins are at a workable depth throughout the district. A number of smaller faults, usually running roughly north-south, also exist, and formerly presented additional difficulties for the collier. It has been estimated that in the Saundersfoot area the total thickness of the coal measures does not exceed 453 yards[1].

Coal Seams

The seams which yielded the bulk of the coal were the Stammers, Rock, Timber, Lower Level and Kilgetty veins. They may be described as follows:

The Stammers Veins were the highest and least important in the Saundersfoot area, and seem to have been worked only to the south of the Harbour fault. Of the six veins in this group (each having a thickness of 18in.-2ft.) four were regarded as inferior, whilst the fourth and sixth veins (sometimes called the Coal and Yard veins) were considered to be of medium quality. These dipped at an angle of up to 36°, and by 1845 had been worked to a depth of 15 fathoms below sea level in the neighbourhood of Saundersfoot Harbour.

The Rock Vein was worked north of the Harbour fault and as far west as Broadmoor. It was of good quality, and in thickness varied from 2ft. 4in at Broadmoor to 4ft. 6in. at Moreton, where at one time an outcrop was worked as a "string". The Low Vein, found in association with the Rock Vein, was of poorer quality, and varied in thickness from a mere 6in.-12in. at Bushmoor to 2ft. 6in at Foxholes.

The Timber Vein had a thickness of about 6ft. when regular, but when it exceeded 6ft. it became very culmy. It was always regarded as soft, producing on average no more than 1 part of large coal to 3 parts of culm. The vein gained its name as a consequence of this softness, because its poor roof required an unusual amount of timber to maintain. Like the Rock Vein, the Timber Vein was worked most extensively in the ground between the Erroxhill and Harbour faults. Some smaller veins of coal, including the Fiddlers and Garland were found beneath the Timber Vein, but these were of inferior thickness and quality.

The Lower Level Vein (sometimes called the Bonvilles Court or Summerhill vein) was found 85-100 fathoms below the Timber

Vein in the vicinity of Saundersfoot village, but extended probably without intermission from the Harbour fault in the south to its outcrop in the north, along a line from Amroth and north of Kilgetty to Thomas Chapel and Reynalton. In the Kilgetty area the Lower Level vein was about 20in.-22in. thick, of very good quality and considered to be well suited for smelting. The vein usually produced 1 part coal to 1 part culm, and although subject to patches of soft coal, the roof was firm and needed little timber.

The Kilgetty Vein (sometimes called the Scad vein) lay about 47-50 fathoms under the Lower Level vein, the interval containing a small and scarcely workable vein called the Catshole, and also several beds of good quality ironstone. Although the Kilgetty vein had an average thickness of just 21in. (and was often only 18in. thick) Messrs. Fosters, writing in 1845, described the quality as "first rate" and "probably the best and purest anthracite in South Wales". The Kilgetty vein produced up to 1 part coal to 1 part culm, an important commercial consideration as coal was easily saleable and culm was not. In spite of their depth and thinness, the extent, quality and regularity of the Lower Level and Kilgetty veins made them the most valuable in the Saundersfoot district.

Iron Ore Measures

Within the coal measures several series of iron ore measures are also to be found, and in the nineteenth century some of these were worked extensively. The most important of these were the Kilvelgy, Fiddlers and Catshole measures. In the Saundersfoot district ironstone was usually referred to as "iron mine" or just "mine".

The Kilvelgy measures lay about 20 fathoms below the Timber vein and 70 fathoms above the Lower Level vein in three workable series. Of these the middle series produced about 9 cwt. of iron mine per square yard, and the upper and lower series about 6 cwt. per square yard.

The Fiddlers measures lay 60 fathoms below the Timber vein, and 30 fathoms above the Lower Level vein. They were very irregular, the pins (or veins) sometimes being so far apart as not to be accessible within the height of an ordinary heading or level. When found in good order they yielded iron mine comparable to the Catshole measures.

The Catshole measures lay 8 fathoms below the Lower Level vein, and produced about 9 cwt. of iron mine per square yard. Catshole iron mine was considered to be good quality iron ore.

Coal Mining Organisation

In the earliest treatise on the Pembrokeshire coalfield George Owen[3] described the coal as occurring in three forms—as an "ore coal", a "string", and a "slatche". "The ore is the best and is a great vayne spreading everyway, and endureth longest; the stringe is a smalle narrow vayne, sometimes two, three or four foote in bignes, and runneth downright, and is alwaies found between two rockes. A slatche they call a peece of cole by itself found in the earth, and is quickly digged about, and noe more to be found of that peece."

The "slatche" appears to have been a small isolated outcrop which was soon mined. The term "string" describes coal as it appeared along its outcrop, perhaps especially when geological disturbance had left the coal near vertical or with a steep angle of dip. The process of digging coal along its outcrop was known as "crop-working", or "smutting". Such mining was usually fairly shallow, and might be effected by shafts or slopes or even by the excavation of a sizeable trench along the outcrop. At Jeffreston, west of Saundersfoot, coal was once worked as a string over a distance of 1¼ miles and traces of the resultant trench may still be seen today north of Jeffreston village.

George Owen's "ore coal" was coal usually won by the conventional means of shaft or slope or level. Writing in 1603 Owen observed that in earlier times pits of four fathoms were regarded as "a great labour", and in any case entrance slopes were preferred to enable people to carry "the coales upon their backes alonge stayres which they called land wayes". Levels were considered an expensive undertaking, costing "sometymes £20 and oftentymes more". In Owen's own day shafts were rarely as much as twenty fathoms deep, the coal and water being raised in barrels by windlass.

Many of the earliest collieries in Pembrokeshire were "bell" pits—large holes in the ground which assumed a bell shape below ground as the miners extracted the coal. When the roof became unsafe a new pit would be opened nearby. Another practice, probably of later date, was to drive a heading out from the bottom of a shaft and take it as far as the ventilation would allow. Another shaft would be sunk to meet the heading and provide fresh air. The colliers could then continue the heading, and when ventilation became a problem once again the same procedure might be repeated. In such circumstances the original air pit might be widened and used for winding. This practice is said to have been used on the Hean Castle estate, and was probably adopted elsewhere. Although it may have been a better method of mining than by the inefficient bell pit, it was still an unsatisfactory way of working coal, and damaging to farm land. By the nineteenth century the Pembrokeshire coalfield was dotted with abandoned, overgrown and usually flooded coal pits. Even then there were complaints about them being left open, creating a hazard to life and limb and a hindrance to agriculture.

The organisation of the collieries was a significant factor in their development. In the Saundersfoot district much of the land belonged to large landowners, some of whom were more interested in receiving rents and royalties than in taking an active part in exploiting the minerals. The result was that from early days pits were usually worked under lease by groups of colliers who were almost invariably short of capital. However at large collieries the lessees sometimes sub-let to smaller teams or companies of colliers, each having some autonomy in their working. Thomas George[4] described such an arrangement in a letter to Edward Loveden in 1808, at the same time estimating how to make a profit on the colliery at Moreton: "If every single company was made up of two men, two women and two boys, their yearly pay at least would be £70. Timber, candles, iron, rope, powder etc. would be little less than £50, say £30, added to £70 makes £100. Six such companies would expend yearly £600, consequently they must sell goods worth £900 at the Pits mouth, if the proprietors are to receive ⅓rds Lords part [Royalty]; Say the proprietor pays to an agent £50 out of his £300—that leaves him a clear profit of £250." Obviously the sole intention of this arrangement was to produce a large profit for the landlord. There is no suggestion that the landlord or coal proprietor should plough money back into the business, and thereby help colliers with new equipment.

Another letter to Edward Loveden, written in March, 1807, by Edward Martin[5], revealed the usual way of assessing the royalty and rent for a colliery lease: "the exact proportion of Royalty or landsmoney should be fixed, and depend on the thickness and position of the veins of coal, the quality of the coal and the contiguity of the colliery to shipping." Martin expressed the view that "the Royalties in Pembrokeshire are very high. Some proprietors or owners of coal land have ⅓, some ¼, and others only ⅙ of what the coal or culm sells for to shipping, after deducting the expense of conveyance or carriage". In fact royalties of just ⅙ were rare, so it was little wonder that coal mining in Pembrokeshire was usually a struggle!

The unwillingness of landlords, and the inability of most lessees, to invest heavily in their collieries was a serious obstacle to the development of deep mining. Although the steam engine for pumping and winding was introduced into South Wales in the 1770's, it was not applied in Pembrokeshire until the turn of the century. With some exceptions, it is generally true to say that technical innovations came to the county late. Although the first half of the nineteenth century saw the formation of several companies to work and develop collieries, the working of small pits by a very few men continued to be a feature of Pembrokeshire coal mining well into the twentieth century. Indeed, when larger concerns closed down, for a while there might be an increase in cropworking, smutting and even the picking over of old coal tips.

Initially a coal mining "company" amounted to little more than a partnership. A company could be incorporated formally by Act of Parliament, as in the case of the Saundersfoot Railway & Harbour Co., but this was rare. The early examples of Saundersfoot coal mining companies—for instance the Tenby & Begelly Coal Co. or Messrs. Williamsons' Merton Coal Co. (at Moreton colliery)— were almost certainly of less formal character. The limited liability company was the creation of the joint stock company legislation of 1855-56, and it seems that the Pembrokeshire Coal & Iron Co., with its considerable coal mining interests, may have been one of the first concerns of its kind to be registered under these Acts.

Broadly speaking, the period of landlords' active participation in South Wales coal mining had passed by the middle of the nineteenth century[6]. In this respect Saundersfoot was not typical. Right up to his death in 1857, Sir Richard Philipps of Picton Castle (Lord Milford after 1847) appears to have been a partner or shareholder in the enterprise at work on the Picton estate. His successors were less involved, but the landowner J. M. Child seems to have had a stake in the Begelly collieries well into the 1860's. Charles Ranken Vickerman was the most notable exception to the general rule. After 1863 he was a substantial landowner in his own right, and he also put up most of the capital required for his collieries and ironworks. The provision of capital was always the key factor in the maintenance and development of the industry. Without C. R. Vickerman it must be doubtful if Saundersfoot's industry would have been able to continue on any scale into the twentieth century.

The Collieries

(a) Shafts

In the eighteenth century the deepest shafts were in the Kilgetty area. In 1780 West Park shaft was sunk 84 yards to the Kilgetty vein, and seven years later the same vein was reached at the same depth at White Gates. Early in the nineteenth century the Thomas Hill pit in Amroth parish was sunk 94 yards, and at some date probably not long after the Engine shaft at Begelly was put down 106 yards to the Timber Vein. The Union pit at Begelly, which was probably sunk in the 1820's, reached the Timber Vein at 120 yards.

Deeper shafts were not sunk until the middle of the nineteenth century. In 1846 the Lower Level vein was found at 168 yards in the Tower pit at Bonvilles Court, whilst in 1854/5 the Kilgetty vein was found at 155 yards at New Hayes, near Thomas Chapel. Between 1853 and 1859 the Grove pit was sunk at considerable expense in the hillside above Kilgetty Ironworks, reaching the Lower Level vein at 132 yards and the Kilgetty vein at 210 yards. Initially there was difficulty in proving the Kilgetty vein, and according to some correspondence of 1859 the shaft was sunk another 70 yards in a fruitless search for the vein. Although coal was not worked at that depth, on this basis the shaft at Grove may be regarded as deeper than any of the workings at Bonvilles Court, where the 'K' pit, an underground slant from the Lower Level vein, was put down to the Kilgetty vein at 242½ yards in 1876/8. Measuring 18ft. × 15ft. the Grove shaft was probably the widest in the Saundersfoot district, and the first to be equipped with cages. In 1873 the Grove pit was actually described as "one of the largest and finest in the Principality", but by then cages were in use not only at Grove, but at Lower Level, Kilgetty and Bonvilles Court. Cages may well have been used at Moreton and elsewhere, but evidence is lacking. It is known, though, that in the late 1860's Moreton's upcast shaft was deepened and put down to the Lower Level vein at 316 yards. So far as can be ascertained this was the deepest colliery shaft in the county.

Colliers always regarded the business of sinking shafts as difficult and specialised work. In Pembrokeshire the task was called "pouncing", and this was described by Franks in 1842 as "another name for boring, taking the means for the end[7]". According to the same account, pouncing was "performed by three women and two men placed opposite to each other, and pressing the ends of two logs acting as levers, which operate on a circular bore". Franks observed that "it was certainly not proper work for females, although it appeared . . . less irksome than the other kinds of labour which women submit to in those parts". Presumably pouncing in this fashion sufficed only to open up the shallower pits: some blasting must have been essential to create the deeper shafts.

(b) Ventilation

The very high carbon content of the anthracite (often 93-96%) usually meant that the coal could not contain unduly large proportions of dangerous gases. Firedamp, composed largely of the very explosive gas methane, was a recognised hazard in the eighteenth century. A 1778 report on the Moreton collieries refers to the problem and implies that inadequate steps were taken to deal with it. Both Begelly and Broadmoor collieries were troubled by it, and indeed Broadmoor (Wilson and Smith's) colliery seems to have gained the reputation of being a fiery pit. On the other hand, Kilgetty colliery, which worked different seams, did not encounter firedamp. The ventilation in this colliery may have been much better. Even so colliers at Kilgetty (and every other pit) had to face some chokedamp (carbon dioxide) as well as the various risks associated with coal dust.

According to R. H. Franks, Inspector for the Children's Employment Commission, 1842, "labour in the collieries of the counties of Glamorgan and Pembroke . . . is, from want of proper ventilation and economy of mining operations, extremely unwholesome, and productive of diseases which have a manifest tendency either to shorten life or reduce the number of years of useful labour . . ." In 1857 H. F. Mackworth, the first inspector of coal mines in South Wales, declared that ventilation in non-fiery pits was still "lamentably neglected[8]". The underground layout of the pits, with narrow roadways, and all the angles created by the pillar and stall system, undoubtedly made ventilation difficult although air shafts were of some use in shallow workings. The practice of using furnaces for ventilation was not very efficient. By this method a furnace was located underground near the upcast shaft, the heat driving the foul air out. At the same time fresh air was sucked in and through the workings from the downcast shaft.

As early as the 1840's mention was made of the occasional use of a form of fan ventilation at Broadmoor old colliery, but the only important use of fan ventilation was at Bonvilles Court. Soon after he took control of this colliery C. R. Vickerman arranged for G. Waddle of Llanelly to instal a large ventilator fan (believed to have been the first of its type) at what became known as the Fan pit. This came into use in October, 1863, and remained in service until the colliery closed in 1930.

(c) Systems of Winding

In the earliest days, as George Owen tells us, winding equipment was not required because pits were entered by a slope or slant, and people carried coal out on their backs. In Owen's own time coal and colliers alike were wound up and down the shaft, usually by four men turning a windlass—apparatus sometimes referred to in Pembrokeshire as a "druke and beam". Coal was carried in a barrel or basket (a corve), whilst the colliers sometimes used a corve and sometimes sat astride a horizontal bar of wood or iron known as the "clatch iron". In the Saundersfoot district the horse gin, or whim, seems to have come into service in the second half of the eighteenth century. It consisted of a drum mounted on a vertical axle placed near but not over the shaft. The drum was turned by yoked horses, causing ropes passing over pulleys to be raised or lowered in the pit. The name Old Engine Pit or Whim Pit for an eighteenth century shaft at Kilgetty probably indicates the use of a horse gin. However the term "engine" can create some confusion. At this time it usually meant a horse gin, although a "water engine" might refer to a water wheel. At the beginning of the nineteenth century water wheels were to be found at Kilgetty, Coombes Colliery, Amroth, and probably elsewhere.

The earliest steam engine in the Pembrokeshire coalfield appears to have been one set up in 1800 at Sir Hugh Owen's Landshipping colliery at a cost of £1,900—a relatively modest sum at the period. Even in 1817 a commentator complained of the lack of steam power in Pembrokeshire, and it seems likely that with the possible exception of steam engines set up at Kingsmoor and Kilgetty, steam power was not used in the Saundersfoot district before the 1820's[19]. Even then there is doubt as to when collieries first adopted steam power: all that can be said with confidence is that before 1820 steam power was unusual in Pembrokeshire, but after 1840 it was widely used.

In the early 1840's Broadmoor colliery and Begelly colliery appear to have made the most use of steam power, their engines producing 75hp and 60hp respectively. At Kilgetty, however, there were three engines producing a total of 56hp, and a water engine (or engines) producing 75hp. There is a dearth of information about the equipment of the collieries at this period, and indeed no further details are known. Suffice to say though that, on the information available, Grove colliery had the largest winding engine in the area, being a Cornish engine with a cylinder diameter of 40in. and 3ft. 6in. stroke.

So far as is known electricity was used only in those collieries at work after 1920. Broom colliery made the most extensive use of this power, approximately 30,000 tons per annum being cut and conveyed by electrical machinery between 1934 and 1939[11].

(d) Drainage

Most of the collieries were drained by adits—levels driven through the hillside to enable water to drain away from the workings by gravity. In the eighteenth century there were at least two drainage adits running inland from the sea at Saundersfoot. One of these, the old Saundersfoot level, appears to have served pits at Ridgeway, Bonvilles Court and in the vicinity of Hill ("Middle Hill") and Pentlepoir. The other level ran almost parallel to this, but further south, serving the Foxholes estate and the pits at Moreton. In the Begelly area there was an adit running west from Spadeland via several small pits to Black Hook and the headwaters of the little Cresswell river. Similarly a number of pits at Kilgetty were drained by a single adit, whilst at Amroth the longest adit of all extended from the coast inland to Coombes colliery before turning north-west to Thomas Hill and Merrixton—a distance of over three-miles!

Until the introduction of steam power the depth and extent of mining was limited to the area effectively served by an adit, or by the capacity of a windlass or horse gin to raise water to the adit level. Even so, in the Saundersfoot district and especially at Kilgetty, the usual practice was to sink shafts some distance to the south of a seam's outcrop to enable colliers to work the seam to the rise. In this way water drained naturally from the workings whenever the seam was above the water table. When a seam was worked to the dip water could be a considerable problem, although in a dry season men might manage to work otherwise inaccessible coal. This point was highlighted in the report on the disaster at Loveston colliery in 1936. Following an inrush of water from old workings, colliers were warned not to underestimate the ability of their predecessors, "the old men", to work coal to the dip.

By 1842 steam pumping was in operation at Hean Castle, Broadmoor, Begelly and probably Moreton. A water wheel provided at least some of the power for pumping at Kilgetty. Grove colliery had easily the largest pumping engine in the district, a machine with a cylinder diameter of no less than 80in. According to one perhaps apocryphal story a new piston for the engine was found to be too large on delivery: several men had to be called hastily to chip it and shave it to size before water flooded the pit!

In 1873 David Parsell, a local engineer, patented apparatus "for pumping, draining and ventilating and applying motive power especially for use underground[13]". The central feature of the invention was a form of vacuum pump which sucked water from mine workings and used it to operate a compressed air engine. A network of pipes then distributed the compressed air around the workings to feed a compressed air locomotive employed in tramming wagons. This system was probably used for a time in Moreton colliery, but does not appear to have had any lasting success.

(e) Lighting

The modern reader, accustomed to electric lighting at the flick of a switch, can scarcely imagine the poverty of colliery lighting even in the nineteenth century. Before the invention of the miner's safety lamp in 1815, colliers everywhere had to rely on candles, flint mills and (rather less dangerous) sunlight reflected by mirrors. Sometimes a faint glimmer might be gained from such oddities as phosphorescent rotting fish. Against this background, Dr. Clanny's safety lamp came as a godsend, not that the light was much stronger. So far as can be ascertained, however, in the Saundersfoot district Clanny lamps, Davy lamps and other safety lamps were used only in the pits at work in the twentieth century. It is possible that the lamps were introduced at Bonvilles Court in the 1870's or 1880's, but it is known that tallow candles held upright in a base of clay were used at Lower Level colliery right up to the time of closure. Good productivity by colliers in these circumstances now seems little short of miraculous.

Methods of Working

This important and somewhat technical subject is not without difficulty, primarily because of the shortage of written evidence relating to mining practice in the Saundersfoot district. References to mining operation are almost invariably brief and very generalised, and although there are still a few men alive who actually worked underground, their very interesting opinions and

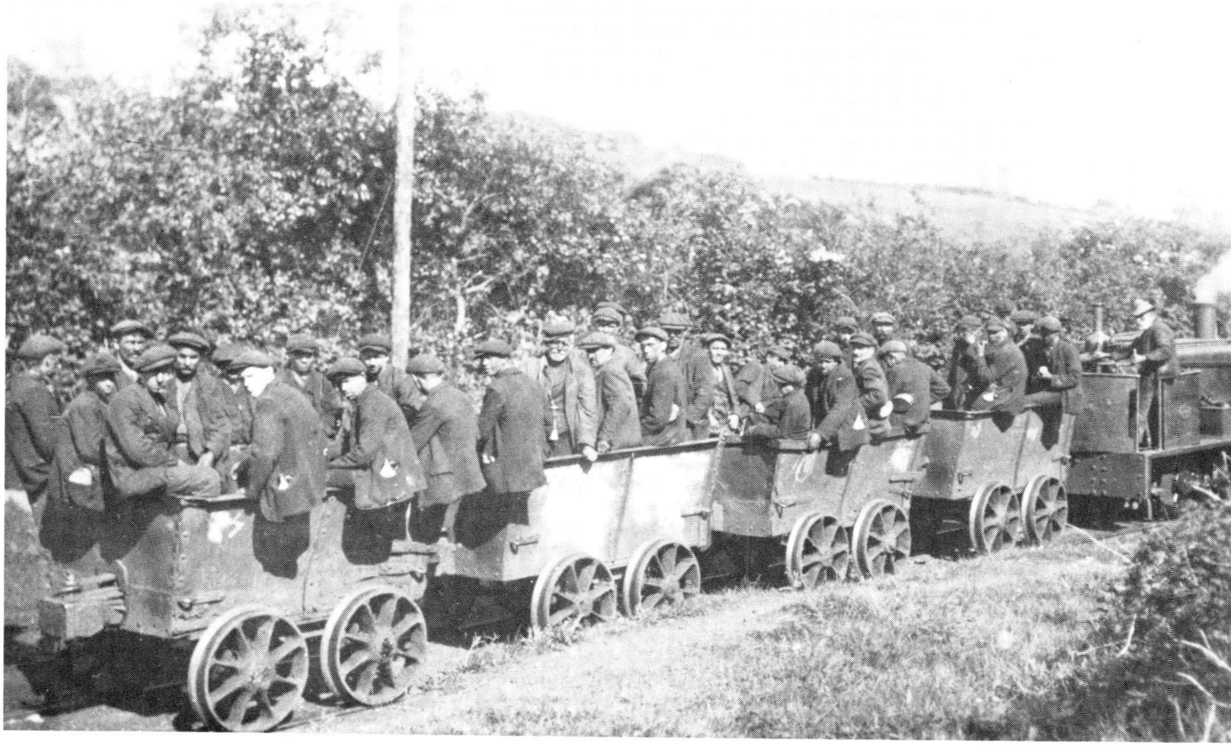

Miners' train at foot of Bonvilles Court incline

"Rosalind" and coal trams at foot of incline

View down incline from Bonvilles Court

Bonvilles Court colliery, 1890's

Bonvilles Court colliery, 1900
(Note the two women at work, lower right)

Bonvilles Court colliery in Edwardian era

recollections do not invariably tally. Accordingly this account is an endeavour to set down information which appears to be reliable, although the reader would do well to assume that any assertion made about practice in one pit at one period does not imply that the same practice was employed in all places at all times.

Broadly speaking there are two main methods of coal mining in Britain, each subject to some local variations. Nowadays the longwall method is customary, whereby coal is obtained from a single working face sometimes hundreds of yards long. The pillar and stall method, once widely used in the principal coalfields, was the method preferred in Pembrokeshire, at least until the twentieth century. This method consists in working portions of a seam ("stalls") in between pillars of coal left to support the roof; this was less wasteful than it might appear because it was customary when retreating from the boundary of the colliery lease to extract the pillars and allow the roof to cave in. Curiously enough pillar and stall working (as opposed to longwall) was usually popular in areas where seams were very thick; this was rarely the case with the deeper seams around Saundersfoot!

In Pembrokeshire variations in mining practice were made necessary by the angle of the anthracite seams, and underground considerable use was made of slopes, slants or driftways. In 1842 R. H. Franks described operations in some detail: "supposing a vein of coal to lie at an angle of 45°, windlasses are fixed at convenient distances on the incline of the vein, by which means (if the mine is worked by adit or level, or above the adit) the coal after being brought from the stalls to the stage of the windlass in carts or skips is dropped by the chain of the windlass down the incline to the level road, and the empty carts are worked up to the stage on which the windlass is fixed by the opposite chain of the windlass. If, on the contrary, the coal is worked to the dip, the coal is in similar manner worked up to a convenient stage by windlass, and then taken by shaft to the surface. These windlasses are worked by women and their labour is certainly severe, though only of 8 or 10 hours duration".

The 1842 report was enlivened by some fascinating illustrations of working procedures, including the unfortunate women working windlasses on "pitching veins". In some mines it appears that these windlasses might be used at several levels underground, so that the section or vertical plan of a mine might be quite complex. However, in the Saundersfoot district, because the angle of dip was to the south, the level mainways usually ran east-west across the dip. Franks mentioned visiting one colliery where the angle of a dip was over 55%; it is believed that there were workings in the Broadmoor area and elsewhere where the coal was actually vertical. In such circumstances the provision of horizontal rather than vertical support became a feature of mining, and usually this made more economical use of timber. As noted earlier, it was usually preferable, for drainage purposes, to work coal to the rise (sometimes referred to as working "landside"). On this basis stalls would be opened at right angles to a heading driven up into the seam on a gradient from a level mainway. However, for reasons of geology or ventilation, or the simple exhaustion of more accessible reserves, it was also often necessary to work to the deep (that is to say, "dipside") following the seam down by driving a slope, off which stalls could again be opened to work more coal.

Cutting the coal was commenced by one of the more experienced colliers lying on his side and removing a thickness of coal from the bottom of the seam with a small pick, or mandrel. Having undercut the coal by a foot or two (depending on the size of the seam) it was supported with timber until the collier was ready to make it fall under its own weight, or with the assistance of wedges or bars[14]. The face worker's task was plainly uncomfortable: he might spend most of his day lying in a confined stall wielding his mandrel to cut coal. Although pads of hard skin soon developed on his shoulders and sides, the cramped and often wet conditions were always unpleasant to work in. A small flat shovel was used to move the coal back in the restricted space of the stall, and another collier (usually called the "butty") then loaded it into a skip or tram for a lad to take out into the heading[15]. As Franks indicated, a windlass or winch would then be used to move the tram to the nearest mainway. In 1842 it was said that a collier would be satisfied if he cut up to 3 square yards in a shift[16]. In later years, when better equipment was available, it is thought that this figure was often exceeded. Even so, the job never got much easier, and perhaps partly because the men were always expected to provide and maintain their own tools. Likewise the collieries never offered such refinements as pit head baths, and a man would have to wait for a wash until he returned to his home[17].

Tramming was one task which did improve in some pits. For efficiency's sake it became desirable to use larger trams in the mainways below ground, and as distances between coal faces and the pit bottom gradually increased the ability of children to do the work diminished. David Parsell experimented with a compressed air locomotive, but pit ponies proved to be more reliable. Accordingly by the 1870's ponies were to be found in a few of the larger collieries. Underground stables for ten ponies or horses were provided at both Bonvilles Court and Grove. Ponies appear to have been a feature of the colliery scene until the closure of Bonvilles Court in 1930. In 1900, however, there were just three horses, one pony and one donkey below ground at Bonvilles Court. By that time steam capstans and winding engines were also in use for moving trams underground. The electrically powered "mane and tail" system of underground haulage is described under the heading of Bonvilles Court in Chapter 3.

In the twentieth century a method of mining known as "open face" was employed at Bonvilles Court and Broom collieries. This seems to have been a variation of longwall mining, although stalls were excavated to create access to what became an extended coal face. In the Bonvilles Court vein (Lower Level or "Low" vein) at Bonvilles Court, stalls were cut 14 yards apart off the heading. Within each stall the men were usually allowed to work 11 yards along the face and 3 yards back, thereby extracting all the coal between the stalls. If the stalls were off a slope to the deep, for example, the men in each stall worked 11 yards "landside" and 3 yards "dipside", until all the coal was excavated from 11 yards landside of the top stall to 3 yards dipside of the bottom stall. Stone which had been blasted down to open up roads and headings was then "gobbed back" into the empty space from which coal had been removed[18].

Opening up these workings required considerable skill, bearing in mind the three dimensional character of a coal mine. An experienced collier was always given the task of "cutting the rib", that is to say, cutting into the coal seam from the top stall at the appropriate angle for it to be mined effectively from all the other stalls to be worked from that slope or heading. Not surprisingly, extra pay accompanied this achievement! At a later stage, as the stalls were opened up and lengthened, a new slope or heading would have to be driven parallel to the first to ensure the good circulation of air across the coal face. Older stalls and spaces might then be bratticed or blocked off for the same reason. Gradually, therefore, the underground plan of the pit would alter as areas were worked out, and either "gobbed up" or sealed off.

After World War I coal mining around Saundersfoot was a curious mixture of the old and the new. Up to the late 1920's there were instances of a few men "smutting" coal, using little more than a pick and shovel to excavate coal, either near its outcrop or through shallow underground workings. Even in the larger pits where more machines were used, traditional methods of mining were not forgotten. However the introduction of electricity improved lighting and working in pits like Bonvilles Court, Broom and Kilgetty, and the extensive application of electricity at Broom gave colliers a glimpse of advantages enjoyed for years in some other parts of the country. Unfortunately the example of Broom colliery was not enough to save Saundersfoot's coal industry.

Colliers

The life of the Pembrokeshire collier was often short and hard. In 1842 R. H. Franks was told by David Morgan, the manager of Broadmoor colliery, that most men were unable to work by the age of 50 or 55. James Bowen, a Narberth surgeon, painted an even grimmer picture. He reckoned the average life of a collier was 40 years, and rarely 45 years. In the Begelly area, where lung diseases were common, he claimed there were not 6 colliers of 60 years in the entire population.

Far more women were employed in the Pembrokeshire coal mines than elsewhere in South Wales. On the evidence of the 1842 report Angela John asserts that the number of adult women then employed in the county was proportionately higher than in any other area employing women in Britain, although there were fewer young girls than in most other areas[19]. Richard Brough, manager of the Begelly colliery, described their duties as "working at windlasses below, wheeling and screening coals above . . ." This was a fairly typical use of female labour. Thirteen years old Ann David hauled skips or tubs with her sister at Begelly colliery, and received payment on piece rates. Usually they worked for eleven or twelve hours a day and pulled about sixty loads. Ann Thomas, aged 17, was employed in pouncing, or pit-sinking, at Kilgetty colliery. She described it as "much harder work than the windlass: it hurts the back".

The hardships of adults were bad enough, but Franks discovered that several collieries around Saundersfoot employed very young children. At Broadmoor the manager said that boys began work at 7 years and girls at 12; Lewis Wilson, the proprietor, simply said that there was no fixed age, but children began work from 7 years. Occasionally a child of 6 might be at work in a mine. James Bowen, the surgeon, took a poor view of it all. He firmly believed that taking children down the mines shortened their lives, and he proposed a lower age limit of 12. Lewis Wilson disagreed, claiming that "a limitation of age would be a barrier to their being brought up to working habits".

The most common task for children in the Pembrokeshire collieries was pulling or pushing coal trams—"tramming". At Broadmoor in 1842 the trams could carry 4 cwt., at Begelly 5¼ cwt., and as a general rule they were moved by no more than two children. In those places where clearances were too low for tramming the children would have to revert to the older custom of hauling sledges, skips or tubs; these contained about 1½ cwt. of coal. Usually one child pushed and one pulled, often using the hated girdle and chain—the rope girdle being worn around the waist and the chain attached to the skip going back between the legs of the boy as he crawled through the workings. At Kilgetty skips were hauled distances varying between 2 and 30 fathoms. At Hean Castle colliery, at least, the girdle and chain was not used, and the skips were hauled with the aid of a wooden handle called a "goff". Generally speaking the children were also responsible for guiding the skips when being raised or lowered on the inclines or slopes underground.

Although the colliery foreman and managers claimed that children worked no more than 8-10 hours each day R. H. Franks found that many rarely worked less than 12 hours, and some 14, 15 or even 16 hours. In all this time there were no regular meal breaks; children just had to eat their bread and cheese and anything else they had, when the opportunity arose. These facts were confirmed by many of the adults interviewed by Franks, who acknowledged that often they worked shorter hours. Sometimes this occurred because the children were not allowed to finish until they had moved a certain quota of coal each day. Richard Hare, agent of the Kilgetty Company, pointed out that the mainways at Kilgetty did not exceed 3ft.-3ft. 6in. in height, and only the young people could work in them. He admitted that although 8-10 hours work each day was usual, children were often down longer because a lot of time was lost in cutting away the heads and ends of rock in the workings to make room. Lewis Wilson of Broadmoor admitted that when the coal was required the youngest children might even work on a shift at night.

R. H. Franks interviewed a number of children from the collieries around Saundersfoot, and most had some simple but telling comment on their working conditions. James Davies, aged 8, worked in a pit owned by Thomas Stokes. In the report it is named as Thomas Chapel, but on other evidence this was surely an error, and the colliery was almost certainly Hean Castle. James Davies said "I have been below one year and I earn 10d a week which my father takes [a typical arrangement] I work with my brother, who is 11, pushing trams. I have never been hurt, and I work longer than my father. I have been to Sunday school but never to day school." Benjamin Thomas, an 8 year old child at Broadmoor colliery, described the work as very hard, and added "my father is dead and my mother works in the colliery with my sister and three brothers. None of the boys in this pit wear shoes". Such were the effects of poverty! William Absalom, aged 13, hauled skips at Kilgetty, working for Ben Howard, a contractor, who paid him in money every other Wednesday. He described the problems of digging a vein 18in.-20in. high, where even crawling was difficult. Lionel Brough, a mining engineer of Neath, told R. H. Franks that the top and bottom of one of the Pembrokeshire veins were "so excessively bad that all the timber they can send down the pit is insufficient to keep the ways good enough for horse-roads". Consequently they were "generally so low and narrow that boys *only* can tram the coal from the hewers to the bottom of the pit." He admitted later that it was technically possible to keep the ways good enough for horses.

Richard Buckby, the Rector of Begelly, stated that children often left school by the age of 10, and that there were no schools in the district for children employed in labour. He added: "if such schools did exist it would be difficult to persuade colliers to send their children to them". The appalling ignorance of the men employed in the collieries of the Begelly district was a point noted by Bourne in 1843[20]. Little wonder that the children learned little! Richard Brough, manager of Begelly colliery, observed somewhat complacently: "The people employed in Pembrokeshire collieries did not labour so hard as English miners; that they were satisfied with small earnings, and made their wages up gardening and agriculture, and few were living about who had not a cow."

The references to gardening and agriculture certainly had some basis in truth. In many ways the Pembrokeshire collier was a countryman, and even in the nineteenth century, in the era of larger and more mechanised pits, the colliers continued to take an interest in agriculture. Mackworth, writing in 1854[21], pointed out that even then many of the pits were shallow and temporary, giving occasional employment to small numbers of people, perhaps only the members of one family. In the circumstances "the Pembrokeshire collier differs but little in any respect from the agricultural labourer, and his gains but little exceed the payment for work on the surface of the ground. He is too poor to move to other localities, such as the valleys of Glamorgan, where wages are much higher; and he has frequently a freehold or other interest in his cottage or hovel". How many actually had a freehold is debatable, but similar points were made in an 1873 Report by Messrs. Lewis & Bidder, who mentioned that cottages and gardens, and up to 4 or 5 acres of land, had been granted to the colliers over many years. "This enabled them to keep their cow and pig, and attached them to their homes. The influence is still felt, by the fact that the colliers' and miners' wages, and the mechanics also, in this district are always and invariably quite 33% or 50% below the standard of other mining districts in South Wales. Sometimes they will leave and "go to the hills", as the phrase is, attracted by higher wages. But they mostly return from Merthyr, Aberdare etc. declaring that they really are no better off for the higher wages, but rather the reverse. The cost of lodging, and the increased cost of food etc. more than counterbalancing the higher pay added to the discomfort and inconvenience of separation from their 'kith and kin'."

Wages were undoubtedly low. In 1806 men were paid[22] 1/- per day, and women 6d.-8d. The mining expert John Curr was astonished, comparing them with rates of between 2/6 and 4/-paid in his part of the North of England. Even so, as George Edwards has pointed out, in Pembrokeshire the collier was better paid than the farm labourer, who might earn a mere 6d a day without victuals. By the standards of modern times pay increases were very occasional and very modest; even in 1835 the men at Begelly colliery received 1/- to 1/6 per day. R. H. Franks' report makes it obvious that the improvements (such as they were) did nothing to diminish poverty. Furthermore although the 1842 report created a stir in Victorian society it did not immediately effect changes in West Wales. The law was reformed, and restrictions placed on the employment of women and children, but such was the shortage of job opportunities in Pembrokeshire that impoverished families had good reason to try to evade the law. This was not very difficult at pits entered by levels or slants, and the fact the mining was often undertaken by sub-contractors made it harder to apportion responsibility and adopt effective controls. Noting that mineral agents sometimes only visited a pit once a month, Angela John maintains that as late as the mid 1850's the law was being evaded in West and South Wales.[23]

In due course the law was made effective. By 1860 children under twelve were banned from the pits and women worked only above ground. By the end of the nineteenth century most pit women were employed screening or sorting coal,[24] although in

Pembrokeshire some were "tip girls", removing cinders from furnaces and unloading waste at the tips. Others collected and prepared timber for pit props. In spite of the speed at which the women were supposed to work there were some lulls, and A. J. Munby sketched one Pembrokeshire woman "knitting between the arrival of cart loads, engaged in a perpetually hopeless task of trying to keep her wool clean as she sat beneath the chutes",[25] The women usually dressed in long flannel skirts, with canvas aprons, headscarves, plaid shawls, black woollen stockings and hob nailed boots. Some tip girls wore "clappers" (loose cloth shields) over the back of their hands to give protection and warmth.[26]

Industrial Relations and Trade Unionsm

The earliest evidence of industrial unrest in the Saundersfoot district appears in the 1850's. The minutes of the Pembrokeshire Iron & Coal Co. for 1857 speak of the need to keep much larger stocks of materials to reduce the effects of any labour troubles. In May, 1859, at a time of depression, the P. I. & C. Co. and the Bonvilles Court Coal Co. actually reduced wages, and threw a large number of men out of work. With many families facing destitution, the men held a meeting on Kingsmoor Common on 16 June, 1859.[27] After some deliberation they marched in order to Merrixton Bottom (Stepaside), Saundersfoot and Bonvilles Court, and on meeting fellow workers still at work "gently reproved them" as the Pembrokeshire Herald put it, "for clandestinely breaking the covenant they had all agreed to". Another meeting was held on Kingsmoor Common a week later, but Captain J. M. Child, who lived nearby at Begelly House, was unconcerned. It was said that Child was "too well acquainted with his men to expect a riot, and there was none".

This reference to the miners' discipline should not be taken to mean that they were timid. Even at this period, such a state of affairs would be unlikely amongst men accustomed to very hard and hazardous employment. The explanation may lie, at least in part, in the fact that many local miners were Primitive Methodists, and firm adherents to the honest and sober values of nonconformity. In the Saundersfoot district employers for the most part seem to have been associated with the established church, and the workmen with the non-conformist chapels. The indications are that here, as elsewhere, Methodism became associated with trade unionism; it certainly coloured the spiritual lives of the coal miners and their families.

Carefully organised trade unionism appears to have begun in the early 1870's. On 24 April, 1873, men employed at local coal and iron works held a meeting at the Miners Arms, Begelly, to form a branch in connection with the Amalgamated Association of Miners. 130 miners declared themselves in favour of the Union.[28] The A. A. M. was already well established in South Wales under the presidency of Thomas Halliday, and in 1874 Halliday himself appeared in Saundersfoot. In May, 1874, the Bonvilles Court Coal & Iron Co. tried to cut wages at Grove and Bonvilles Court collieries. The men met on Kingsmoor Common and decided to ask the management to let them resume work for a fortnight at the old rates of pay. According to the Tenby Observer, William Foley (for the management) "with a kindness and forebearance that has distinguished all his dealings with the men, met them in a most amicable manner," and he agreed.[29] Soon after Thomas Halliday arrived and won a small concession, but the men would not accept the terms. Instead they attempted to gain a closed shop policy in the company's collieries in return for acceptance of a 5% reduction in wages.[30] The attempt failed, but the most serious setback affected everyone: during the strike the furnaces at the ironworks were taken out of blast. Few guessed then that they would never be re-lit.

The lessons of 1874 were not well learned by either side. In 1876, when the depression was deeper than ever, further wage cuts provoked a much longer strike and an even greater reluctance to compromise. The local press, which usually favoured the employer, now showed more sympathy for the strikers. Pointing out that rents, and the cost of provisions, had not been cut, the Haverfordwest & Milford Haven Telegraph declared in October, 1876, that wage reductions "have gone as far as they ought". The B. B. C. & I. Co. refused to give ground, and went into liquidation at the end of the year. The outcome was not only desperate poverty for hundreds in and around Saundersfoot, but the destruction of the miners union. However as trade revived self-help revived, and in 1881 the Monmouthshire & South Wales Miners Provident Society started a Pembrokeshire branch.[31] A year later it had 334 members.

The disastrous 1870's had a noticeable effect on wages and population. After 1875 wages in the coal industry were customarily paid on a sliding scale, regulated by the selling price of coal. To operate this system a joint committee of coal owners and workers was set up and maintained by both sides.[32] The arrangement was supposed to protect the men against arbitrary action on the part of the employers, but in spite of various amendments to the sliding scale it also tended to stifle effective trade unionism. The results were not very beneficial. In 1881 1/6 per day was an average wage for a workman at Bonvilles Court; in 1873 it had been 2/- to 2/6. In 1873 colliers could earn as much as 3/6 per day, but they did not recover such rates before the end of the century.[33] A pay sheet from Grove and Lower Level in March, 1897, shows that boilersmiths and engineers were amongst the highest paid, the latter receiving 2/7 per day. A woman trammer of 1897 earned 1/- per day![34] Not surprisingly the last quarter of the century saw a decline in population. In 1861 the total population of the parishes of Amroth, St. Issells, Begelly and East Williamston was 4,222; in 1881 the figure was 3,752.[35] Low wages, or the lack of employment undoubtedly encouraged emigration, both within the British Isles and beyond. It is recorded that in the summer of 1871 twenty-five Primitive Methodists emigrated to America and elsewhere "owing to the Kingsmoor coalmasters ceasing to work their pits."[36]

The lot of the Pembrokeshire collier did not improve greatly in the twentieth century. Although the 1898 strike in the coal industry did not have a very serious effect in Pembrokeshire, in South Wales as a whole it indicated deep dissatisfaction with the sliding scale agreement, and paved the way for the setting up of the South Wales Federation of Miners. From its formation in October, 1898, the Federation grew rapidly, and Saundersfoot became the district with the smallest membership to be represented on the executive council.[37] In 1912 the strength and solidarity of the union was put to the test in the miners demand for a fair minimum wage. The demand was nationwide, and the response wholehearted. On March 1st the historic dispute began: in the words of R. Page Arnot "the strike was of a magnitude that had never occurred in any single industry in any country of the world".[38] In this great movement the men of Saundersfoot were only a tiny part, but it was soon reported that "the miners to the number of 350 were amongst the first to leave work in the country."[39] The end of the strike five weeks later was a relief for all concerned, but the eventual outcome for the Pembrokeshire men was a good deal less than satisfactory. A Minimum Wage Act became law, but Pembrokeshire coal owners were exempted from the standard applied elsewhere in South Wales. Not for the first time, thin seams, difficult working conditions and the reputed low cost of living in Pembrokeshire were used to justify the payment of lower wages.[40] These arguments were not without some truth, however, and if the Glamorgan standard had been applied in Pembrokeshire collieries might well have closed.[41]

During World War I control of coal mines was taken over by the government, a move which may have been in the national interest at the time, but which also sowed seeds for later argument. Despite the 1919 recommendation of the Sankey Commission that the industry should be nationalised and the men given higher wages the government decided to return the mines to the coal owners. Few of them had the resources either to pay more or to invest heavily in machinery to improve productivity. The result was the 1921 Miners' strike which had a severe impact on the Saundersfoot district. The three month dispute produced no winners, and did little more than sour already difficult relations between mine owners and men. Further trouble was always likely, and it came in 1926 as soon as the government withdrew certain subsidies from the industry. A long and fruitless strike brought bitterness and hardship to every coal mining district. In Saundersfoot some men went without pay from April to November, and the suffering they and their families endured was considerable. It was a measure of how little had changed that four soup kitchens were opened in Saundersfoot during the strike of 1926.[42] Shades of the 1870's, when the Rev. J. H. A. Philipps distributed beef and flannels to the destitute! As usual on these occasions colliers searched for coal on the beaches or old spoil tips, and their families eked out an existence until the strike was over. Fortunately for some, the traditional link with agriculture remained strong, and provided some support. In 1930 it was estimated that smallholdings were held by 55% of local colliers.[43] After the closure of Bonvilles Court that year a number of men found work at Hook colliery, near Johnston. In 1939, after the closure of Kilgetty and Broom collieries, 224 men joined the unemployed register at Tenby, but over half were soon involved in wartime construction work.[44] Union attempts to resist these closures proved to be in vain.

Footnotes

[1]G. S. W. C. provides a comprehensive account of the geology of the district.

[2]PC 4099, Vickerman, G. S. W. C. The colliers sometimes referred to the Peacock Vein, on account of the colours and highlights to be seen in anthracite from this source. It is not clear whether this was a feature of one or more of the coal seams, but the name may have related to coal from the Timber Vein. Most samples of Pembrokeshire anthracite now available display a hard brightness, but not always with such highlights.

[3]George Owen (1603) "The Descripton of Pembrokeshire" Cymmrodorion Record Series, I.

[4]Goggerdan, Letter of T. George, dated 11.3.1808.

[5]Goggerdan, Letter of E. Martin, dated 19.11.1807.

[6]Morris & Williams, p. 116.

[7]Children's emplopyment commission (1842).

[8]H. F. Mackworth, Inspector of Mines, Report (1857) quoted by Morris & Williams, p. 189.

[9]Goggerdan, Letter of H. Jeffreys, dated 29.3.1807.

[10]Cambrian Register (1817) quoted by Edwards (1963), p. 12. The National Library of Wales holds a copy of an Ordnance survey preliminary survey, dating from the period 1811-1820, which shows a steam engine on Kingsmoor to the west of the Begelly-Tenby road, and some distance south of Begelly church. As regards the Kilgetty engine, see the section headed Kilgetty Colliery in Chapter 4.

[11]Edwards (1963), p. 14. Although electricity was not used to the same extent at Bonvilles Court, the colliery is said to have had a sufficient surplus of power to provide electric lighting for some of the houses in the village!

[12]NCB/R13541.

[13]Patent No. 1596/1873.

[14]In areas where coal seams were thicker than in south Pembrokeshire it might be necessary to undercut the coal by two or three feet. The process was known as "holing the caol". As seams were thin and the ratio of coal to culm often 1:1, colliers seem to have been cautious about bringing coal down by shot-firing. However shot-firing was used a good deal to break down rock and waste material. A rock drill used in preparation for shot-firing at Bonvilles Court was included in the "Saundersfoot's Industrial Past" exhibition opened at Kilgetty Information Centre, 1976.

[15]As the years went by, two sizes of underground tram were introduced in some pits, the smaller ones being used near the coal face and the bigger ones in the mainways.

[16]Children's employment commision. Evidence of S. Singleton to R. H. Franks.

[17]At work colliers usually wore a flannel shirt, trousers and boots—although even in the twentieth century men sometimes worked barefoot. In addition to tools they usually carried their lamp, food tin, water jack (flask) tobacco box, timber scribing knife and maybe a cartridge pricker.

[18]The writer has been told when the pillar and stall method was employed at Bonvilles Court stalls were 12 yards apart. Howells includes a glossary of local dialect, including some coal mining terms.

[19]"By the Sweat of their Brow: Women Workers in Victorian Coal Mines." A. V. John (Croom Helm, 1980) p. 38.

[20]M. A. Bourne "Guide to Tenby" (1843) 15, 59. Carmarthen.

[21]H. F. Mackworth, Inspector of Mines Report (1854) quoted by Edwards (1963) p. 30.

[22]Goggerdan, Report of J. Curr to E. Loveden (1806) quoted by Edwards (1963) p. 28.

[23]"By the Sweat of their Brow: Women Workers in Victorian Coal Mines" p. 57.

[24]In a letter to a local newspaper in 1926 a collier mentioned six grades or sizes of coal, namely beans, nuts, large coal, peas, rubble and duff.

[25]"By the Sweat of their Brow: Women Workers in Victorian Coal Mines." A. V. John (1980) p. 89.

[26]"By the Sweat of their Brow: Women Workers in Victorian Coal Mines." A. V. John (1980) p. 184.

[27]P. H. 24.6.1859.

[28]T. O. 24.4.1873.

[29]T. O. 11.6.1874.

[30]T. O. 30.4.1874 reported on activity in the Pembrokeshire coalfield, and declared "It is obvious that Pembrokeshire is pretty much a law unto itself as regards wages."

[31]T. O. -.7.1882.

[32]The Chairman of the Sliding Scale Committee was Sir William T. Lewis, later Lord Merthyr, and owner of Hean Castle, Saundersfoot.

[33]Vickerman: Messrs. Lewis & Bidder's Report, 1.2.1873. The report notes that "after recent rises" the wages for colliers were 3/6 per day, masons 3/- to 3/6, carpenters 3/- to 3/4, labourers 2/- to 2/6.

[34]Displayed at "Saundersfoot's Industrial Past" exhibition, Kilgetty Information Office.

[35]Edwards (1950).

[36]Primitive Methodist Minute Books of the Pembroke Dock Station, County Record Office, Haverfordwest. I am indebted to David Howell for this reference.

[37]"South Wales Miners" (A History of the South Wales Miners Federation, 1898-1914) R. Page Arnot (G. Allen & Unwin, 1967) General Appendix 1, p. 379.

[38]"South Wales Miners", R. Page Arnot, p. 290.

[39]H. M. H. T. 6.3.1912.

[40]"South Wales Miners", R. Page Arnot, Appendix to Chapter 11, p. 331.

[41]This was claimed by Walter Roch, a local Liberal M.P.—H. M. N. T., 29.5.1912.

[42]W. T. 13.5.1926.

[43]Edwards (1963) p. 30.

[44]Edwards (1950).

Chapter Three

THE COAL INDUSTRY: COLLIERIES SOUTH OF THE ERROXHILL FAULT

The task of providing short histories of the collieries is not without difficulties. As has been seen already, the eighteenth century concept of a colliery was rather different to that of the late nineteenth century, and comparisons can be misleading. Furthermore, even before the eighteenth century was out, the Saundersfoot district was riddled with shallow workings, most of which went entirely unrecorded and have long since vanished. Although the nineteenth century saw the development of larger and deeper pits, the lack of records makes it far from easy to distinguish between those which were significant and those which were not. Only towards the end of the nineteenth century do the records offer a more detailed account of the collieries then at work, a state of affairs due partly to the legal requirement to preserve plans of mine workings, and partly to the fortunate survival of a quantity of private papers.

Three other complications must be mentioned. First the title of a colliery frequently referred to different pits at different dates. The name Broadmoor colliery, for example, was attached to at least four different workings within a century. Secondly a colliery title often referred to a group of small pits worked under the same management. This occurred at Kilgetty, Thomas Chapel, Moreton and probably elsewhere. Finally, as the years went by, it appears that colliery leases were granted which overlapped with areas leased in earlier generations. In most cases the coal leased was probably in a lower seam to that previously worked, but in other instances it would appear that the same coal may have been worked from different collieries. The most notable example relates to the anthracite between Moreton and Broadmoor. In the early 1870's the proprietors of Moreton colliery took over leases formerly held by Broadmoor, but after the closure of Moreton in

1887 the workings of Bonvilles Court colliery extended into ground previously mined from Moreton and Broadmoor.

In the circumstances there is no very neat way of recording the history of each pit, but a geological and geographical approach appears most convenient, distinguishing between the collieries to the south of the Erroxhill fault and the collieries to the north of it. Mining south of the fault was virtually finished before the end of the nineteenth century, the only very important exception (Bonvilles Court) being attributable in part to the fact that this colliery was the only one in this area of the coalfield to work the deep Kilgetty vein. This chapter deals with the pits south of the fault.

Stammers This colliery was not only south of the Erroxhill fault, but south of the Harbour fault as well. Unfortunately most of its career is shrouded in mystery. It seems to have comprised shafts and crop workings (now largely obscured) at and inland from Swallowtree Bay. According to Fosters' Report of 1845 six veins of coal were found in this vicinity, the most notable being two of medium quality known as the Coal vein and the Yard vein. By this time they had been worked to a depth of 15 fathoms below sea level, in workings to the east of the Swallowtree stream in Rhode Wood (Road Wood).

In view of its proximity to the sea, Stammers may well have had its heyday in the eighteenth century, before the harbour was built. However there are a few references to the colliery in the mid-nineteenth century, suggesting that some mining was maintained for many years in an area which had already been worked extensively. In 1853 it was noted that the Coal and[1] Yard veins, together with six veins of iron mine, were being worked both by shafts in Rhode Wood and by levels into the cliff. The iron mine ground was that sometimes referred to as "Saundersfoot". In the same year Bridges Harvey (who was C. R. Vickerman's brother-in-law) obtained a mineral lease of the land from Richard, Lord Milford. The colliery was at work in 1857 because Mr. Tregelles was taken to see it during his inspection on behalf of the Pembrokeshire Iron & Coal Co.[2] (see Chapter Five). Although a fresh lease of the property was granted in 1873, Lewis and Bidder that year referred to the veins at Stammers as being out of use but

Bonvilles Court colliery

Bonvilles Court miners ''the Boers''

Pay Day at Bonvilles Court; miners outside wages office

Miners train at Saundersfoot, 1909 (Perhaps best known picture of this subject)

workable. It seems most likely that coal mining at Stammers ceased in the 1860's.

Trevane, Hopshill, Hollybush and Vatson These localities lying to the south of the Harbour fault witnessed small scale coal mining in the eighteenth and early nineteenth centuries. In 1778 a Mr. Hopkin was in charge of workings four fathoms deep and two hundred yards long at Trevane.[3] The coal vein, only 9"-18" thick, lay on land belonging to Lewis Pryse. By 1789 Thomas Roberts was in occupation of the coal works here and at Vatson, and accounting to Edward Loveden.[4] At the same period coal was being worked on Edward Loveden's land at Hopshill, and in July 1808, Hopshill Farm and colliery were sold together[5]—an interesting instance of the association between coal and agriculture in Pembrokeshire. In 1792 Holly Bush colliery was in the hands of Messrs. Knethell & Williams, who were also in partnership at the Picton's Moor colliery[6] at Wooden. By 1807 Holly Bush was worked by William Thomas & Co.[7]

Moreton The title Moreton Colliery seems to have applied to different workings at different dates, but all lay just to the north of the Harbour fault. To judge from the earliest references, in the 1750's the landowner Lewis Pryse made an agreement allowing a neighbour, Sir John Philipps, Bt., to work the mines at Moreton.[8] In December, 1776, Sir John's successor, Richard, Lord Milford, was employing forty people at Moreton,[9] and in September, 1778, Joseph Woolstenhulme described the Timber Vein at Moreton as "the best vein of coal that is now worked in Pembrokeshire, producing 160 skips per day".[10]

After the death of Lewis Pryse responsibility for the estate passed to Edward Loveden, who had married Pryse's daughter Margaret in 1773. Loveden and Lord Milford soon quarrelled. The argument seems to have begun with Woolstenhulme's report on the state of Moreton collieries in 1778. Loveden attempted to tighten up on the terms of Lord Milford's lease, and enforce payment of the one-third royalty on coal and culm raised. Lord Milford objected, but in eloquent and forthright correspondence Loveden accused him of inconsistency, injury to the works and (by the way!) giving offence to his family. Milford denied almost everything, saying he was totally ignorant of the particulars of the complaint and that any inconsistency "originated from your conduct in permitting every rascal who solicited your (coal) works at the head of the level to benefit by the labour and expense of my Father and myself who are got many hundreds out of pocket by your works."[11] Not surprisingly the rift between the two men became irreparable, and Milford moved out of Moreton soon after.

Edward Loveden was not lacking in drive or ambition. He was once described as more like "a substantial yeoman rather than a person of education and condition, but he did not want plain common sense, nor language in which to clothe his ideas."[12] It is some measure of his abilities that he became M.P. for Abingdon from 1783 to 1796, and owner of the substantial estate at Buscot Park, near Lechlade. He was also prepared to move with the times and to consider the latest inovations. Accordingly he contacted John Curr, a leading mining engineer and the man usually credited with the invention of the plateway. Replying from Sheffield on 12 May, 1797, Curr wrote "I am of opinion your colliery at Tenby if opened in a proper manner and drained by a Fire Engine [Steam Engine] should afford you a . . . Rent of ¼ of the sale of the coal, the undertakers paying all expenses of opening, and think also the adventurers ought to have a lease of 25 or 30 years." Not forgetting his own interests, Curr added "You will by some means I hope obtain a Way leave, and the undertakers will lay a Cast Iron Rail Way upon the construction I have invented . . . and which is now in great use in sundry parts of the Kingdom. Your idea of running the coals into the vessells (as they do at Newcastle upon Tyne) is a very good one if it does not materially injure the coals by breaking them."[13] At this date Tenby was the nearest settlement of any size to Moreton. However the "Cast Iron Railway" was almost certainly intended to run to the nearest point on the coast at Saundersfoot.

So far as is known none of Curr's ideas were acted upon, although in 1806 Loveden consulted him again. John Curr then described the Moreton collieries as "mere scratchings on old workings"[14] which should have been abandoned a century earlier. In spite of this dismissive observation, he repeated his suggestion for a railway to the shore at Saundersfoot. Conflicting

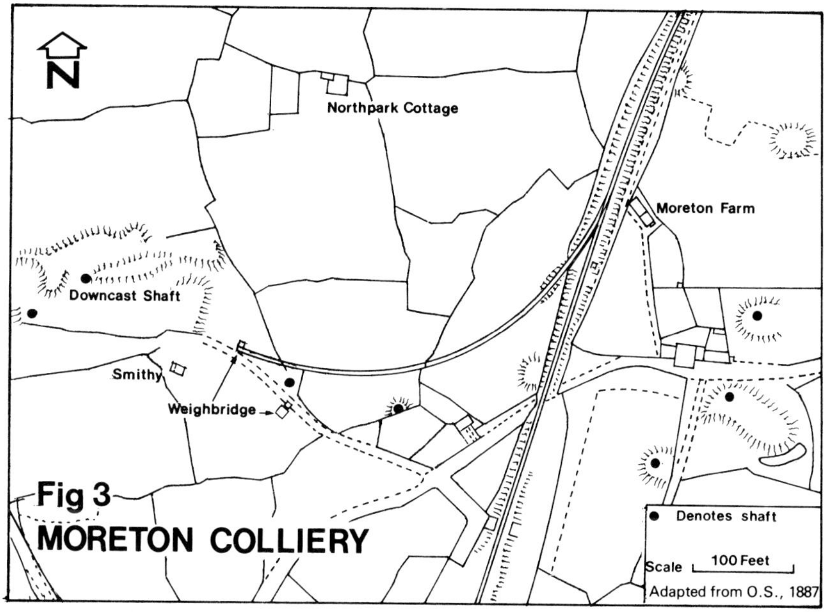
N
Northpark Cottage
Moreton Farm
Downcast Shaft
Smithy
Weighbridge
Fig 3
MORETON COLLIERY
Denotes shaft
Scale 100 Feet
Adapted from O.S., 1887

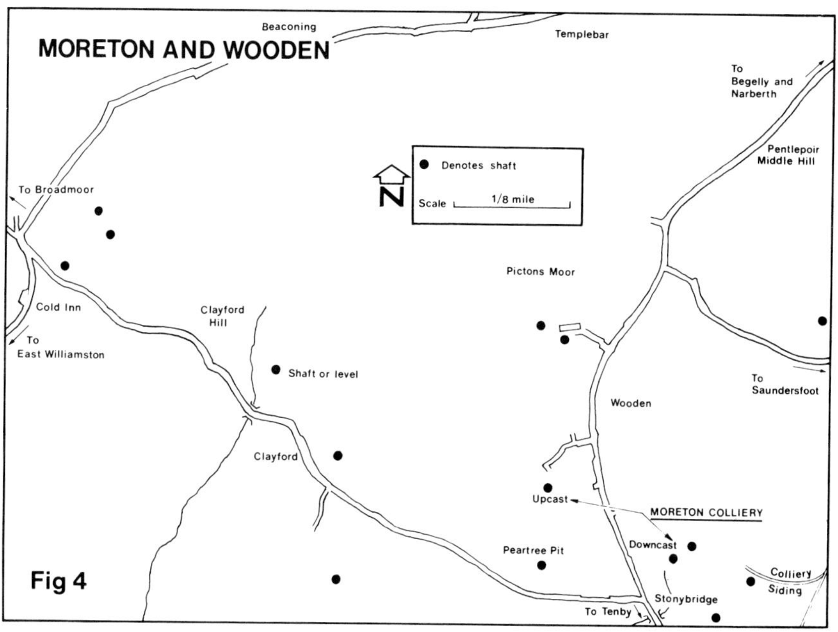
MORETON AND WOODEN
Beaconing
Templebar
To Begelly and Narberth
Denotes shaft
N
Scale 1/8 mile
Pentlepoir Middle Hill
To Broadmoor
Pictons Moor
Cold Inn
To East Williamston
Clayford Hill
Shaft or level
To Saundersfoot
Wooden
Clayford
Upcast
MORETON COLLIERY
Downcast
Peartree Pit
Colliery Siding
Stonybridge
To Tenby
Fig 4

advice was in fact a feature of this period. Owen Thomas, a tenant on the Moreton estate was pessimistic, thinking that a steam engine would not help because little coal remained in the area to be drained by it.[15] Humphrey Jeffreys, on the other hand, saw clearly that the lack of effective drainage prevented the exploration or working of deeper veins. In 1808 he noted that "the pits at present at work (and these are three if such they may be called) are about 16 yards deep—the working of them cannot be carried to any extent for they are only getting small parcels of coal which have been left here and there by former miners, and all the coal which is now above water in the estate cannot be considered as worth notice."[16]

If Jeffreys and Curr agreed upon the poor state of the Moreton collieries, Jeffreys advice was the more positive. He proposed a trial boring on the south west side of the estate, but recognised that the dip of the coal might be a problem. He also advocated the use of a steam engine to raise water to one of two existing drainage adits, and recommended a suitable site to permit coal to be mined in the adjacent Hopshill and Moysland estates. Other advisers broadly agreed with Jeffreys. Edward Martin thought there was more coal west of Stonybridge, near Moreton, and suggested the extension of adit levels to give better drainage.[17] Thomas George thought that this would be too costly, but was convinced that more coal would be found if a trial was made west of Moreton. Like Martin, he offered some advice on the re-letting of Moreton colliery.[18]

Unfortunately the outcome of all these recommendations is not known, although it seems probable that the works were improved and re-let. Whether steam power was provided is doubtful: no further information has come to light concerning Moreton in the next twenty years. The year 1828, however, saw the start of another chapter in the colliery's history. Three landowners, Sir William Barlow, George Lort Phillips and Henry Leach, then agreed to grant a three year mining lease of coal and culm under Holly Bush Farm, East Williamston, Stoney Bridge and Wooden—areas all close to Moreton itself. The new lessees were John Pryse Eyton, Edward Eyton, Samuel Williamson and Septimus Stephen Williamson.[19] As if in recognition of this new development the plans prepared for the Saundersfoot Railway in November, 1828, included a proposed branch line from Saundersfoot to Merton, or Moreton. The lessees, who must have had their lease extended at a later date, soon began trading as the Merton Coal Co.

During the 1830's the two Eytons seem to have dropped out of the business, but on 31 December, 1838, Pryse Pryse (successor in title to Edward Loveden) granted the Williamsons a 31 year lease to work coal and culm under 600 acres of land at Moreton, North and South Vatson, Kitehill and Hopshill.[20] A period of colliery reconstruction ensued, but it was 1 June, 1840, before it was noted that the pit was coming into active work. By that date the Merton Co. had shipped three or four cargoes of culm.[21] As the branch railway was still unbuilt, its construction was advocated once again. Once more the S. R. & H. Co. approved the plan in principle. Unfortunately they were embarassed by Messrs. Williamsons' wish to build on a different alignment to that authorised—evidently to avoid an incline, and to save money. Lengthy negotiations made no progress. The Moreton—Saundersfoot line, although the district's earliest idea for a railway, was never built.

After a seemingly hesitant start, by the mid-1840's Moreton settled down to a period of steady production. The difficult economic climate of the 1850's may well have made an impact, however, because in 1860 Moreton was said to be "limited in extent, shallow and working only a few tons per day".[22] On the other hand, in April, 1862, the colliery was described was "well managed by J. Williamson and his "Man Friday" J. Thomas".[23] In 1864 Thomas Davies, agent to Pryse Pryse, was advised by a mineral surveyor, G. Daniel, that coal was getting short and the grant of a new lease was desirable.[24] At the time the workings were in the Rock and Timber veins at a depth of about 120 yards. Subsequently, the upcast shaft at Moreton was put down to the Lower Level vein at 316 yards. As the downcast was not deepened it is possible that the upcast was also enlarged and bratticed, or divided. The investment was presumably undertaken on the strength of the grant of a new lease; at least one was granted to Samuel Williamson in 1869.[25]

It is a pity that the story is so obscure, because there is evidence to suggest that Moreton was the scene of some interesting experiments at this period. A local engineer and inventor, David Parsell, collaborated closely with Septimus Stephen Williamson of Moreton colliery. At least three of Parsell's inventions were patented, and in two cases the patent was granted to Parsell and S. S. Williamson jointly. One patent dealt with condensing apparatus for steam engines,[26] and another dealt with making more efficient use of steam by passing it from one cylinder to another to work the piston of each.[27] Unfortunately it is not certain whether Parsell applied these ideas in locomotives or only in stationary steam engines, but there is good reason to believe that at least one locomotive application was attempted. In April, 1874, the local press described the first run of a "trucking machine" at Moreton colliery: the design was said to have been patented by David Parsell, the "George Stephenson of the Saundersfoot district." Parsell's third patent already mentioned in Chapter Two undoubtedly did have motive power implications. To judge from its text there can be little doubt that this invention, including the compressed air locomotive, did operate for a while in the workings of Moreton colliery.

Parsell's "trucking machine" was put to work shunting wagons on the standard gauge Moreton colliery siding; described as a "powerful but light and elegant machine", on at least one occasion it made a return trip over the Pembroke & Tenby Railway to Tenby.[28] The Whitland extension of the P. & T. R. opened in September, 1866, and initially Moreton was the only pit in the district to be favoured by the provision of a siding direct to the colliery. Until 1 July, 1870, the P. & T. R. was worked by the contractors, Messrs. Davies & Roberts, but after the company assumed responsibility for running its trains Messrs. Williamsons became responsible for the colliery siding, including the lighting of lamps and the locking of points on the line.[29] This was a matter of some importance because on 29 September, 1869, an accident occurred at the junction of the siding with the main line. An excursion from Shrewsbury comprising "9 waggons and some 800 passengers" was overdue, but the Saundersfoot stationmaster left the signals at danger and went to superintend the shunting of coal trucks from the colliery. Six loaded wagons were standing on the main line, and the "donkey engine" had gone into the siding for more when the excursion ran into them at 25-30 m.p.h. Two detached wagons took most of the impact, and fortunately no one was hurt.[30]

In August, 1872, the Moreton and Trevayne estates were sold at auction to C. R. Vickerman, who thus became one of Messrs. Williamsons' several landlords.[31] The pattern of land ownership around Moreton was complex, and in the nineteenth century the colliery worked under leases from the Picton estate, the Leach family, the Lort Phillips family and others. However C. R. Vickerman's acquisition was significant in that it gave him an interest in the largest colliery in the district hitherto outside his influence. Before long he was exchanging letters with S. S. Williamson over arrears of rent. What bearing this had on subsequent events cannot be determined now, but by April, 1874, plans were afoot to put Moreton colliery under the control of a new limited company.[32] This was the Saundersfoot & Tenby Collieries Co. Ltd., in which at the outset Charles Hinde and George Blizard appear to have been principal shareholders. The Williamsons retained an interest, nevertheless, and in 1876 took a mining lease in their own names for the benefit of the colliery.[33] By then the severe slump in the coal trade was biting deeply, and may have discouraged Messrs. Hinde and Blizard. No reference to them has been found after January, 1876. Thereafter Messrs. Williamsons took charge of the company, which extended its activities to Broadmoor and Jeffreston.

The economic depression soon took its toll, and after a struggle to pay its way the S. & T. C. Co. Ltd. was forced into liquidation in 1879 on the petition of William Frankish and other debenture holders.[34] A. A. James of 110 Cannon Street, London, an accountant in partnership with H. J. Edwards, was appointed liquidator and receiver, but he enabled the Williamsons to continue at work—no doubt because they were the people most likely to obtain some return on capital invested in the colliery. Arguments with C. R. Vickerman over rent arrears persisted until 1880, by which time he had had the remaining minerals in his Moreton and Trevayne estates valued at £28,000.[35] By 1882

Moreton colliery was once again controlled by Messrs. Williamsons, working on a 31 year lease for a rent of £250 per annum.[36] Mining of the Rock vein ended in the same year, and of the Lower Level vein soon after.[37] Most of the underground working was now due west of the upcast shaft in the direction of East Williamston, and there was litle or no mining east of the downcast shaft. The Timber vein appears to have been worked right up to the closure of the colliery on 30 April, 1887. The lease was then turned over to C. R. Vickerman and the colliery partially dismantled.[38] In 1900 the colliery was said to comprise two shafts to the Rock and Timber veins, each of about 50 fathoms (100 yards) and one shaft to the Lower Level vein of about 150 fathoms or 300 yards. At this date the engine houses were still standing, and one 14″ × 24″ stroke double winding engine with nearly new drum and winding gear was still in situ.[39] Eighty years later there was very little to be seen. The spoil tips were cleared after World War II, leaving only small mounds of waste at the sites of the upcast and downcast shafts. Now few holidaymakers passing by on the main road are ever aware of Moreton's industrial history.

Stonybridge, Middle Hill, Williamston and Clayford Small coal mines operated at a number of sites near Moreton in the eighteenth and nineteenth centuries, and it is possible that for a time some of them may have been worked as part of the Moreton colliery. This may not have been the case with the Carew colliery which was to be found at Wooden in 1756,[40] but Stonybridge lay little more than a quarter of a mile south-west of Moreton's downcast shaft on Lewis Pryse's land, and in 1778 coal was worked in this vicinity by David Thomas.[41] In 1792 Stonybridge colliery was in the care of the Executors of the late Henry Phelps. In the same year Middle Hill colliery, Pentlepoir, was operated by one William Jenkins,[42] whilst the nearby Pictons Moor colliery was in the hands of William Knethell, William Davies and Isaac Lewis.[43] The royalty at Pictons Moor was paid to Edward Loveden.

To the west of Moreton, Williamston and Williamston Meadow collieries were worked by Alexander Smith, William Bowen & Co. in the period 1792/93.[44] Williamston Meadow was still occupied by William Bowen & Co. at the turn of the century but by 1809 Williamston was worked by Thomas Murray & Co.[45] In the same year South Field colliery was in the hands of Thomas Whittan and William Thomas.[46] A couple of years earlier the Hill Moor colliery was worked by William Thomas[47] & Co. At least some of the output of these pits seems to have been shipped from Saundersfoot.

These collieries closed with the introduction of deeper mining and better technology. The S. R. & H. Co. thought Stonybridge important enough to propose it as the terminus of a branch line in 1841, but there is evidence for some small scale mining nearby later in the nineteenth century. In about 1870 Messrs. Jenkins & Morgan had a surface working to the east of Moreton's downcast shaft, and another running east-west to the north of the upcast shaft.[48] As late as 1889, J. F. Lort Phillips and Henry Leach made an agreement with William Hilling on working minerals under land at Clayford and Clayford Park.[49] This enabled Hilling to open a small pit which later (in his view) contributed to a dispute between himself and C. R. Vickerman. It is thought that the pit closed after a very few years but there may well have been other examples of such workings in this neighbourhood.

Masterlands The Broadmoor area probably witnessed a considerable amount of crop-working in the eighteenth century and perhaps earlier. However the earliest records so far found relate to the colliery at Masterlands. On 21 January, 1791, Richard, Lord Milford granted a 21 year lease of the pit to Isaac Lewis, William Waters and Alexander Smith.[50] By 1792 the same partners appear to have been working neighbouring ground held by the executors of the late Henry Phelps.[51] Mining probably continued on some scale for many years, because in the early days of the Saundersfoot Railway there was a proposal for a branch line to Broadmoor and Masterlands. Although plans were prepared, and the line authorised in 1842, the idea came to nothing. In 1863 property at Masterlands was occupied by S. Roper and William Harlow; the former was probably the same Mr. Roper as the one then involved with Broadmoor colliery.[52] The workings at Masterlands comprised a slant and at least one shaft, reaching the Timber Vein at 32 yards.[53] Coal was last mined here before World War I, although these operations may have been no more than smutting. Five men are believed to have been employed at Masterlands in 1905. In 1910 Arthur Boncis of London was granted a 42 year lease to work mines under Masterlands, but if work ever began it was short-lived.[54]

Broadmoor (Wilsons & Smith) The story of this colliery probably begins in the 1820's, but the earliest known reference to it is under the name of "Mr. Wilson's colliery" in 1833, at which time the colliery was established well enough to attract the interest of the proprietors of the Saundersfoot Railway.[55] At this period relatively little Broadmoor coal was dispatched through Saundersfoot; most seems to have gone via Cresswell Quay on the Cleddau, a comparable distance away to the west. By 1840 the colliery was known as Wilson & Smith's, and in the same year Abraham Leach granted a 21 year lease of coal veins under East Williamston to Lewis Wilson and James Wilson of Cresswell, and Alexander Smith of Jeffreyston.[56] Mining rights may well have been obtained from other local landlords, including the Carew estate.[57] According to R. H. Franks, in 1841 the colliery had steam-power of 75 h.p., water and winding engines. The main shaft was said to be 34 fathoms (68 yards) deep; the only other reference indicates that the Timber Vein was reached at 14 yards. Wilson's Pit, as this shaft was sometimes called, may have been linked to the nearby Waters Pit but evidence is lacking. The colliery was probably closed by 1854 on the expiration of the existing leases, or in anticipation of the opening of the new Broadmoor colliery. Today the site is marked by some tree covered spoil tips, and one or two small ruined buildings.

Broadmoor (Greenhill) In or about 1850 James Spencer of Jeffreyston and Capel Coape of Putney, near London, collaborated in the development of a new colliery to the south-west of Masterlands. This was the Greenhill pit, the main shaft of which was sunk 194 yards to the Timber Vein. In 1851 Richard, Lord Milford granted the partnership a 21 year lease of coal and culm mines under Morgans Lane and Water Hill, Broadmoor,[58] and it is thought that a lease or leases must have been obtained from George Lort Phillips, another notable local landowner. The colliery is believed to have opened in 1853,[59] and by the 1860's trading appears to have been conducted by the Broadmoor

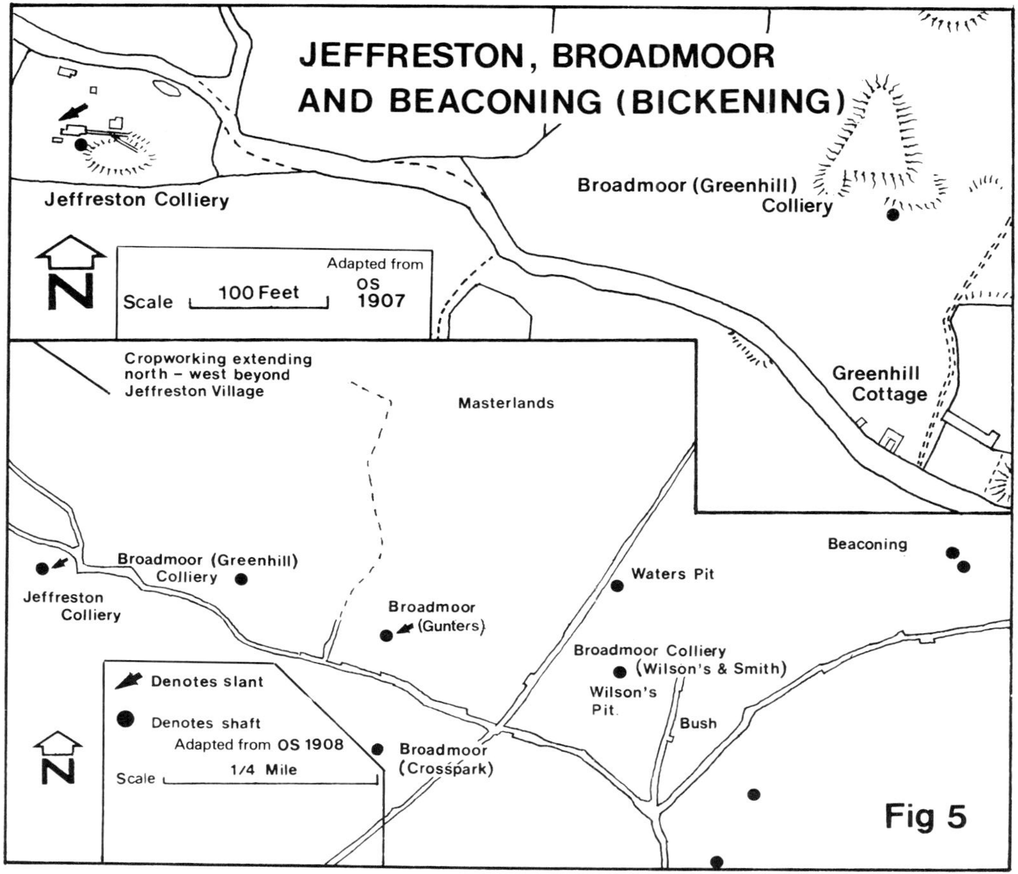

JEFFRESTON, BROADMOOR
AND BEACONING (BICKENING)
Jeffreston Colliery
Broadmoor (Greenhill)
Colliery
Greenhill
Cottage
N
Scale
100 Feet
Adapted from
OS
1907
Cropworking extending
north – west beyond
Jeffreston Village
Masterlands
Beaconing
Waters Pit
Broadmoor (Greenhill)
Colliery
Jeffreston
Colliery
Broadmoor
(Gunters)
Broadmoor Colliery
(Wilson's & Smith)
Wilson's
Pit
Bush
Denotes slant
Denotes shaft
Adapted from OS 1908
Scale
1/4 Mile
N
Broadmoor
(Crosspark)

Fig 5

Anthracite Coal Co. with Adam Murray as Clerk and Samuel Roper as Manager.[60] What part James Spencer played in the business is not clear, but it may be significant that in 1867 Mr. Roper was advertising Jeffreyston House as available to let.[61]

In 1868 a new 21 year lease was granted by the Rev. J. H. A. Philipps to James Coape of Niton, Isle of Wight, but prosperity evidently did not follow. In 1870 the royalty payable on coal was reduced from 1/- per ton to 9d. per ton.[62] In the same year and in 1872 James Coape negotiated new 31 year leases from another landlord, Henry Leach.[63] The land involved lay on the north-west side of the road from Templebar Farm to East Williamston, in the area previously worked by Wilsons & Smith. Meanwhile, in 1871, Messrs. Williamsons of Moreton colliery agreed to purchase some of the Broadmoor colliery leases,[64] an indication perhaps that operations at Broadmoor were beset by financial difficulties. At all events in 1875 James Coape sold his entire interest in Broadmoor colliery to Messrs. Williamson.[65] On 3 February, 1876, Henry Leach gave S. S. Williamson a new 31 year lease at Broadmoor,[66] and the 1872 lease was made over in Williamson's favour.

The Williamson's takeover appears to have put an end to the Broadmoor Anthracite Coal Co. and to have put the colliery into the care of the Saundersfoot & Tenby Colliery Co. Ltd.[67] The failure of the S. & T. C. Co. in 1879 signalled the demise of the Greenhill pit. Although Messrs. Williamson personally recovered control of assets held by the company they preferred to concentrate their efforts at Moreton. In September 1880 they gave notice of their intention to determine Broadmoor's Picton estate lease on 29 September, 1881.[68] There is no sign that Broadmoor (Greenhill) was worked after that date, and it may have shut down earlier. However after the bombing of the Llanion oil tanks at Pembroke Dock in 1941 it is said that large quantities of waste oil were disposed of down the Greenhill pit. The shaft has since been capped, and the site cleared. In the 1970's even the spoil tip was removed, although a short row of cottages built nearby by the colliery company are still occupied.

Jeffreston George Owen mentions that mining was taking place in Jeffreston parish in 1603, but this may have been centred in the

Cresswell and Cresselly areas, close to the river Cleddau. The history of mining near thc village of Jeffreston is obscure, but it is likely that coal was dug in this vicinity by the eighteenth century. Cropworking was prevalent, as is indicated on old maps by the excavation of a long trench running across the north side of Jeffreston village on the north rising of the Timber vein. Even today some shallow workings survive in this area, although heavily overgrown.

Jeffreston colliery, apparently one and the same as Meads Colliery, seems to have been opened in the 1870's and at any rate by 1878.[69] There is some slight evidence to suggest that this pit, like Broadmoor (Greenhill) may have been held briefly by the Saundersfoot & Tenby Colliery Co. Ltd.[70] Whilst this is not certain, the development of the pit may have been an attempt to keep the coal industry alive locally when the Greenhill pit was nearing closure. Whatever the truth of the matter, the signs are that Jeffreston colliery was neither large nor flourishing. Its main feature was a slant driven into the south rising of the Timber vein where it was inclined to the south at no less than 80°., thus appearing as a near vertical ''string''. At a depth of 40 fathoms (80 yards) the vein was found to become horizontal and rise to the north. So far as is known this seam was not worked to the rise.[71] By the turn of the century Jeffreston colliery was in decline. It had 22 men in 1902, 12 in 1906 and only 2 in 1910[71]—but even so it seems to have survived at least until 1913. It is said that the spoil tips were picked over by colliers during the 1921 and 1926 strikes, but the site has since been cleared.

Langdon At some uncertain early date shallow workings were excavated in the fields south of Langdon Farm, but the last of the pits lay almost due east of Langdon to the north of the lane from Langdon to Rose Park. This colliery on the Timber vein is thought to have been opened in about 1906. It comprised a slant facing west, with an air pit only 22 feet deep just to the east of it. The strata dipped to the south by 18″ in the yard. In 1910 seven men were employed here, but production probably ceased in the same year.[73] There is no trace of this colliery today.

Broadmoor (Crosspark) Very little is known of the earliest workings on this site, but it seems likely that they ended in the first half of the nineteenth century. Originally there was at least one shaft, but the coal may also have been worked in a trench as a "string". The Timber vein is said to have been vertical to a depth of 36 yards in this vicinity. Mining at Crosspark resumed after World War I and ended again in 1927. Surviving plans under the heading "Broadmoor Colliery" indicate that the pit had a slant and an air shaft but that the workings were shallow. A collier who once worked at Crosspark has recalled that access was by a shaft about 70 feet deep, and that water was pumped up to an adit level at a depth of 36 feet. The coal seam was very steep, and the pit invariably wet. In 1926 water broke into the workings and one man had a remarkable escape from drowning by being trapped in a pocket of air.[74]

Broadmoor (Gunters) This pit was in the small village of Broadmoor, behind cottages on the north side of the Broadmoor-Jeffreston road. It was opened in 1926 by the partnership of Gunter & Brinn, but was soon under the sole direction of W. C. Gunter. In 1927 six men were employed below ground and two above. A surviving plan shows a slant or drift descending 51 feet, with an engine house alongside the drift mouth. An upcast shaft just 22 feet deep lay a short distance to the west. These very small workings in the Timber vein were abandoned in September, 1930, but local inhabitants can still remember coal leaving the pit by lorry or horse and cart, mostly for local sale.[75]

Ridgeway and Griffithston The area between Moreton and Saundersfoot was dominated after the 1840's by Bonvilles Court colliery. Prior to that decade the area was dotted with shallow coal pits. The most important appears to have been Ridgeway, which was being managed in 1776 and 1777 by Lord Milford, who employed only about 7 miners.[76] In 1812 Richard, Lord Milford granted a 21 year lease of coal and culm mines at Saundersfoot, Upper and Lower Ridgeway, Frogwell Bank, Foxenholes and Little Griffithston to William Hughes (of Killewen), a farmer, and four colliers, Thomas Lloyd, George Phelps, William Hughes (of Hill) and John Lloyd.[77] These men presumably worked the mines for some years, although nothing is known of their activities. The original proposals of the S. R. & H. Co. included a short branch line at Ridgeway, but this never materialised. Coal mining at Ridgeway appears to have finished in the 1830's, although the shafts could be seen for the rest of the century. For many years, until after World War I, the village's piped water supply came from a disused coal working near Griffithston Farm.[78] A number of these old pits were linked by an underground drainage adit which emerged on Saundersfoot beach.

Bonvilles Court Bonvilles Court colliery was founded in 1840 by J. P. M. Myers in partnership then or later with J. Townsend Kirkwood.[79] The first shaft (the Landing Pit, later known as the Fan Pit) was sited to the east of the premises known as Bonvilles Court, and at an early date the workings seem to have connected underground with a heading from the nearby Cants colliery on the Hean Castle estate. Production began by the end of 1843, and in February 1844 the S. R. & H. Co. noted, without enthusiasm, that Myers was sending out speculative consignments of coal. It was thought that this would end co-operation between coal shippers, but in fact it opened up a valuable new market. By June 1844, Bonvilles Court coal had been sent to Rochester "being the first time I believe that anthracite has been applied to the purpose of hop-drying". (J. Longbourne[80]). Within three years Myers & Co. had become the largest shippers at Saundersfoot. Building on this success, a new shaft was developed in 1846.[81] Situated near the old fortified mansion known as Bonvilles Castle, it was named the Tower Pit. Production increased, and in June, 1849, it was reported that the colliery had a stack of at least 7,000 tons of coal (exclusive of culm) which, with the usual output, would probably be shipped that summer.[82]

In or before 1849 J. M. Child granted Myers' company, the Bonville's Court Coal Co., a lease to work anthracite under his land, excluding the house called Bonvilles Court, then occupied by John Longbourne. Child soon suspected that coal was being worked beneath the premises, and commissioned a survey from J. Jones. He was refused access to the colliery, but somehow managed to go down the pit about a month later, accompanied by two assistants. They discovered that the workings ran under the prohibited area, and advised Child accordingly.[83] Regrettably the outcome is not recorded, but presumably Myers was rebuked. The dispute cannot have been too serious, because in 1857

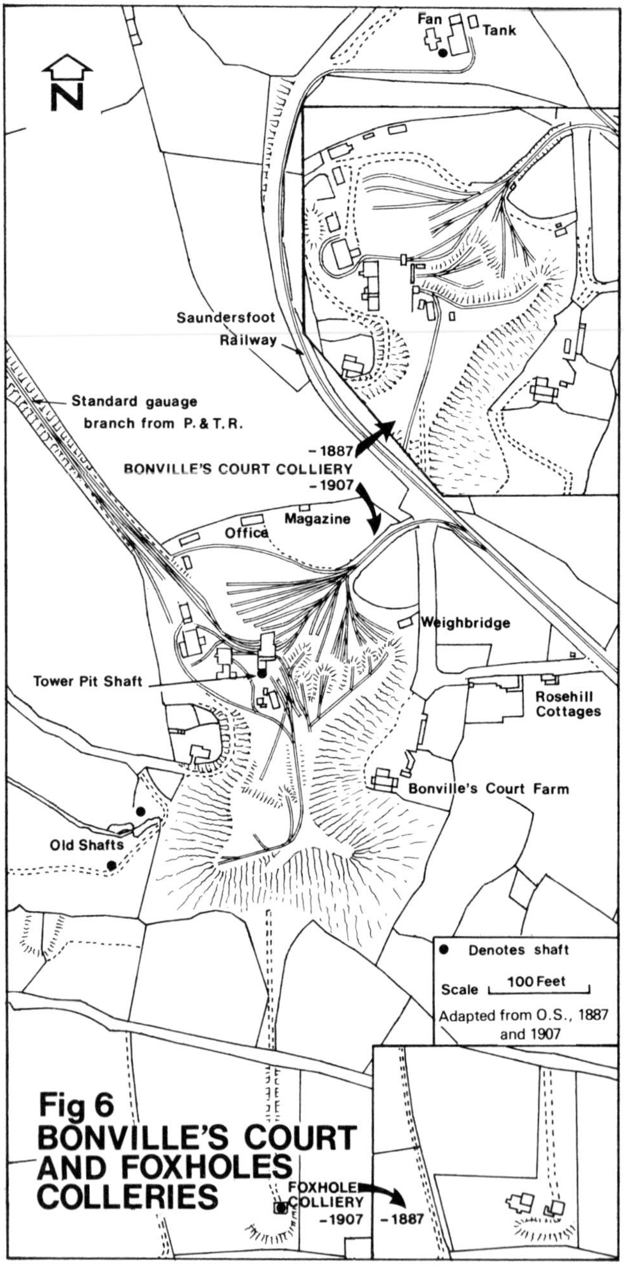

Fig 6
BONVILLE'S COURT
AND FOXHOLES
COLLIERIES

Mouth of tunnel between Railway Street and Coppet Hall

Coppet Hall, 1900's (Former Hean Castle colliery site on hillside behind house)

The middle or ''short'' tunnel on the railway at Coppet Hall

Railway below the cliffs at Hean Castle

Miners digging coal on the beach during 1921 strike

The brickworks and foundry adjoining the railway, Wiseman's Bridge

Foundry workers, Wiseman's Bridge. Group including John and Tom David (centre and right)

The beach at Wiseman's Bridge, looking towards the Inn, the Patches and Amroth

Kirkwood obtained a wayleave from J. M. Child, who owned much of the surrounding land.[84] Myers and Kirkwood together had leased coal and clum from Edward Wilson in 1855.[85]

The available figures suggest that in the years between 1857 and 1861 Myers enjoyed good business.[86] After April 1862 however there is no record of Myers involvement, and by 1863 the colliery was in the hands of C. R. Vickerman. He had plans for the concern, and may have consulted S. H. Blackwell, who reported on Kilgetty ironworks. In any event in October, 1863, a Waddle ventilator fan was installed at what became known as the Fan pit; the ventilator may have been the first of its type ever constructed.[87] It seems likely that at the same time a Patent fuel works was built at Bonvilles Court. Lewis & Bidder (1873) record that the plant comprised an engine, boilers, steam hammer, moulds and the other requisites for the manufacture of patent fuel from culm. Notes dated 1882 say that good quality patent fuel was made, but such a large proportion of materials had to be imported that the works produced insufficient profit. By 1882 the steam hammer had been sold.[88]

When the Bonvilles Court Coal & Iron Co. was established in 1873, the company prospectus declared that the Bonvilles Court collieries were well known throughout England for the highly priced "Myers Malting Anthracite coal", especially suitable for the preparation of best malt. Myers enterprise in 1844 plainly served to perpetuate his name! (The prospectus is reproduced in Appendix No. 1). The brief era of the B. C. C. & I. Co. saw further developments at the colliery. In the 1870's both shafts received cages and guides. The Fan shaft (measuring 10′ × 9′) was equipped with a single cage and the Tower shaft (measuring 11′ × 10′) had a double cage. The former shaft went down 120 yards to the Lower Level vein, and the latter reached the vein at 168 yards. Between 1876 and 1878 the 'K' pit was sunk to the Kilgetty vein, 242 yards deep. The unusual feature of this shaft was that the pithead was entirely underground, about 110 yards west of the Tower shaft. Colliers who recall the arrangement say that the wheels and winding gear, engine and boilers were all contained in a cavernous space underground. The 'K' pit shaft was 11′ in diameter; alternative access to the Kilgetty vein workings was provided by a slope 114 yards long on a gradent of 1 in 2½.

Forty years of energetic mining inevitably ate into anthracite reserves, and by 1881 the workings had reached Saundersfoot Bay in one direction and a fault known as the Newton fault in the other. Concerned about future prospects C. R. Vickerman decided to risk money on sinking a shaft at Foxholes, just to the south of Bonvilles Court. The story of that colliery will be told shortly, but Vickerman still wanted to maintain the colliery at Bonvilles Court. The problem was to negotiate new leases on terms good enough to justify the endeavour. C. R. Vickerman was an eloquent correspondent, however, and not above a little exaggeration in arguing his own case. On 30 August, 1881, he wrote to J. M. Child's successor, A. T. Hook Child, asserting that the upper seams at Bonvilles Court were almost worked out, and that for some time he had been thinking of closing the colliery and concentrating on production at Grove and Kilgetty. He described Bonvilles Court as "a losing concern". About a year later he told Hook Child that they had found sufficient coal to last two years, and in October, 1884, he suggested that without a new agreement on rents, royalties and wayleaves the life of Bonvilles Court could not be estimated at more than a year. Hook Child was not rushed into a helpful response by the possible threat to his income. It was not until 1888 that he granted a new 21 year lease.[89] In the meantime, presumably under pressure of circumstances, C. R. Vickerman allowed his men to encroach on coal belonging to J. Harcourt Powell, for which he had previously failed to negotiate a lease. Powell got wind of this development in 1883, but Vickerman admitted responsibility. Joseph Cadman, who had worked at Bonvilles Court in the 1860's as C. R. Vickerman's agent and manager, carried out an inspection on Powell's behalf.[90] It was not until February, 1887, that the ensuing litigation was settled in Powell's favour.[91] J. Harcourt Powell cannot have had any lasting grievance because in 1891 he leased the property, known as the Bickening lands, to C. R. Vickerman. In the same year Vickerman obtained at least five other leases, mostly at East Williamston, in an area worked earlier from the collieries at Moreton and Broadmoor. Minerals at Pentlepoir were leased in 1896, and in 1900 a new lease was obtained in respect of 1600 acres of the Picton estate.[92] Together these leases transformed the outlook for Bonvilles Court at a time when the few other pits were closing down.

At the turn of the century Bonvilles Court was easily the largest colliery in Pembrokeshire, and the only one of any significance in the Saundersfoot district. Appendix No. 2 records the most important pieces of equipment at the pit in 1900. For all its importance, Bonvilles Court was small by the standards of South Wales. Nevertheless after the closure of Lower Level it provided employment for about three hundred people, and made a vital contribution to the local economy. Although many colliers continued to have a small holding or some other interest in agriculture poverty was never far distant. In a sense this one colliery became a bulwark against destitution. Unfortunately many papers were lost in the fire at the offices of the Bonvilles Court Coal Co. in 1913,[93] and even more recent records have disappeared, but it seems that the period prior to World War I was a time of steady output and adequate profitability. After the peak year of 1914, when approximately 42,000 tons of coal were produced, output fell slightly with a drop in manpower, perhaps on account of the war.

In its later years Bonvilles Court was, by Pembrokeshire standards, almost an ideal colliery. Well ventilated, well maintained and for the most part dry, working conditions for the collier were considered to be good in spite of the narrow seams. The chief problem was that after eighty years of continuous production many of the coal faces were about three miles from the pit bottom. More than one collier is said to have walked from Broadmoor to the colliery to go to work, and then walked from the pit bottom almost back to Broadmoor in order to reach the coal face! This state of affairs inevitably created difficulties in removing anthracite from the pit. The solution provided was an underground haulage system depending on two electric engines, one above ground and one below. They operated an arrangement known as "mane and tail", involving the action of two separate drums, one winding the haulage rope and the other unwinding it, the rope being guided through the workings by sheeve wheels and passing around a stationary wheel at the far end. One end of rope was attached to the front of the first tram and the other to the back the last one, so making an endless line. As many as thirty trams ("the journey") might be moved by this method at any one time, the first electic engine taking the trams out as far as a parting a quarter of a mile below ground, whilst the No. 2 engine (located above ground) moved them to the pit bottom ready for winding up the shaft in the cages. The "mane and tail" system operated with the skilful assistance of a "haulier", a man who rode behind the last tram either going in or coming out, and who used his feet to push the rope into the sheeve wheels on each corner and change of gradient.[94]

In spite of the "mane and tail" system, the extent of the underground workings, and the dwindling reserves of workable coal posed a threat to the future of the colliery. A new engine and chimney stack was constructed near the Tower pithead in 1927 when Bonvilles Court was managed by J. C. Samuel, but the outlook was not rosy.[95] Miners strikes in 1921 and 1926 probably weakened the company, and by 1929 it had to cope with all the consequences of falling output. In July, 1929, the prospects had changed to such an extent that the management sought a 10% reduction in the men's wages.[96] The men protested, claiming that the fall in output was not their fault, but a consequence of having to work in exceptionally confined "stalls" underground. The men were given notice to take effect from 3 August, 1929, and closure was thought to be imminent. Steamers crowded the harbour to collect the available anthracite, but negotiations were successful and Bonvilles Court was back in production before the end of the month.[97] This proved to be no more than a reprieve. Anthracite was increasingly difficult to mine, and a report from Jasper G. Daniel, a mineral surveyor of Swansea, indicated that there was no prospect of extending the workings.[98] Notice of closure was given, and the last shift was worked on 17 April, 1930. Some 320 men lost their jobs, although about 40 were retained until June 4 to carry out dismantling work underground.[99] As late as October 1930, it was suggested that a syndicate planned to form a new company to work minerals formerly worked by the Bonvilles Court Coal Co., but nothing materialised.[100]

The closure of the colliery did not put an end to all activity at the Tower pithead. Most of the buildings remained intact, and when Broom colliery re-opened new coal screens were constructed to handle coal brought down on the Saundersfoot Railway. For most of the 1930's traffic was worked as required on the standard gauge siding to Bonvilles Court, originally opened in 1893. The end of mining at Broom in 1939 spelt the end for the screens at Bonvilles Court, and by 1940 most of the equipment had been removed for scrap. During World War II some of the spoil tips provided hardcore for roads and runways, but the colliery site was left to decay. The area of the Fan pit became more and more overgrown, but in the 1950's and 1960's the site of the Tower pit was tidied up and made suitable for holiday caravans. A few buildings survive today, together with the mound marking the old pithead, but probably few visitors appreciate how important Bonvilles Court once was to Saundersfoot.

Foxholes C. R. Vickerman developed Foxholes colliery primarily to break out of the strait jacket of old agreements made with A. Hook Child and perhaps other landlords. In July and August, 1882, he described the enterprise as a "trial pit". No doubt past experience caused him to be cautious. Even so, he invested a good deal of time and money in sinking the new shaft, and the work was still in progress a year later.[101] A corrugated iron engine house was built on a stone foundation, and a large boiler installed. An incline, described as a "slope tramway" was constructed on the surface to convey coal from the pithead up to the screens at Bonvilles Court. The design of the incline was based on that at Grove colliery, Stepaside. When completed in 1884 Foxholes had a shaft 9′ in diameter going down 120 yards to the Timber vein and also intersecting the Rock vein. Rather curiously, for this late date, there is no indication that the shaft was either divided or duplicated, although the pit may have been connected underground to the workings of Bonvilles Court. In any event it was not a success, and by 1887 the shaft was disused. The engine house and incline were still in existence at the turn of the century, but are believed to have been dismantled soon after.[102]

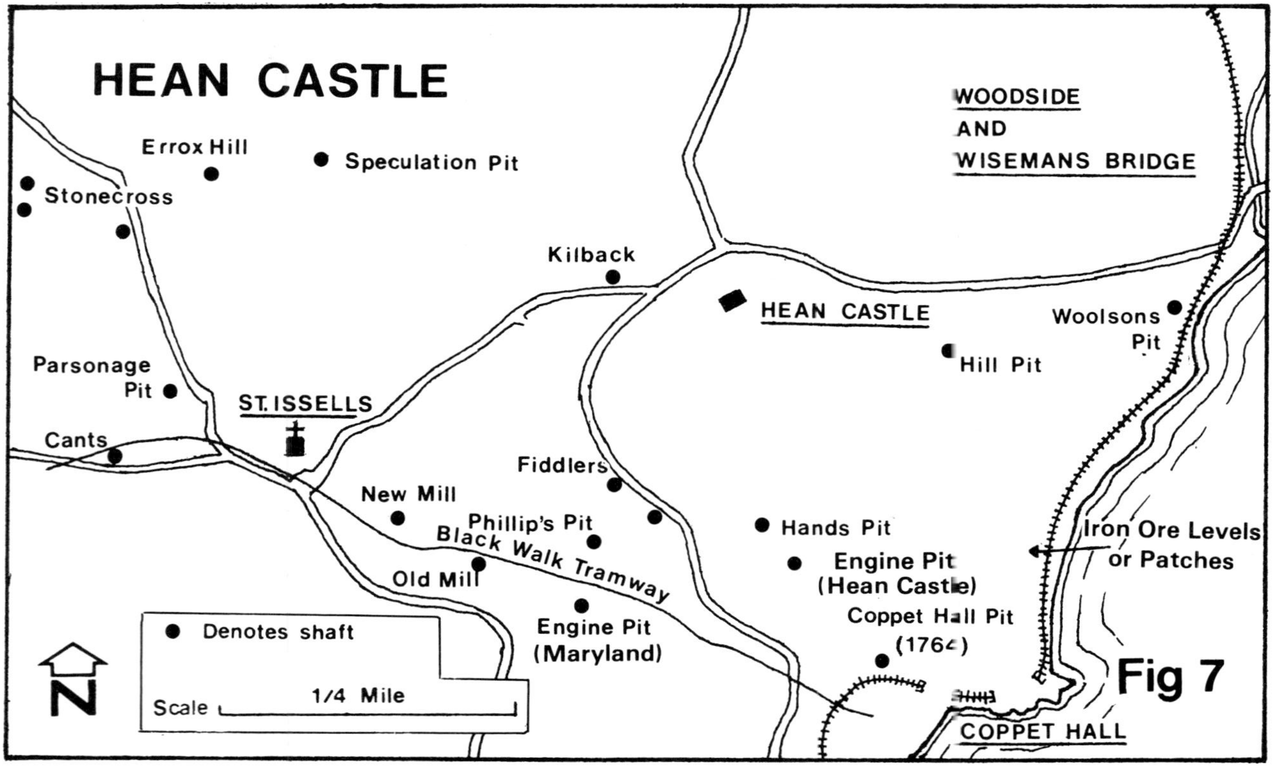
HEAN CASTLE
Errox Hill
Speculation Pit
Stonecross
Kilback
WOODSIDE
AND
WISEMANS BRIDGE
HEAN CASTLE
Woolsons
Pit
Hill Pit
Parsonage
Pit
ST. ISSELLS
Cants
Fiddlers
New Mill
Phillip's Pit
Black Walk Tramway
Old Mill
Hands Pit
Engine Pit
(Hean Castle)
Iron Ore Levels
or Patches
Engine Pit
(Maryland)
Coppet Hall Pit
Denotes shaft
N
Scale
1/4 Mile
COPPET HALL

Fig 7

Hean Castle (Coppet Hall and Cants) Coal mining on the Hean Castle estate was at its height in the period 1760-1860, but the known facts about the numerous pits are so limited that it is difficult to paint a clear picture of developments. Workings in the eighteenth century seem to have been small scale and widely distributed around the estate. As early as 1764 there was a colliery close to the shore at Coppet Hall,[103] and it is known that by 1786 John Wogan, the then estate owner, was working pits at Stonecross and St. Issells.[104] After 1800 the estate passed into the hands of the Stokes family, and it is probable that the active pits were less plentiful but more productive. An important area of mining at this time was the triangle of land around St. Issels church, ranging from the Kilback shaft in the north to Cants colliery in the west and Maryland in the east, where an engine pit was only about 350 yards from Coppet Hall beach. These pits were linked to the beach by what became known as the "Black Walk tramway" which, although extremely obscure, may be attributed either to John Wogan or to the Stokes family by virtue of being built entirely on Hean Castle land.[105] Some of the pits may have been in the care of lessees, but it seems likely that those closest to Coppet Hall may have been connected underground and worked as one colliery—almost certainly the Hean Castle colliery referred to in records of the 1830's and 1840's.

References to the Stokes family are contradictory, but Thomas Stokes evidently inherited the Hean Castle estate on the death of his father in 1840, his mother retaining an interest. Thomas Stokes was certainly active in the early 1840's, in spite of a depression in the coal trade. In 1843 it was reported that the colliery agent had been travelling for orders and that the pit had a large stock of coal and culm.[106] R. H. Franks visited the colliery in about 1841, although he seems to have confused it with Thomas Chapel in writing up his report. On the assumption that Thomas Stokes evidence given under the heading of Thomas Chapel actually related to Hean Castle the colliery was then not more than 38 fathoms (76 yards) deep, and used a steam engine (or engines) of 46 h.p. for winding and raising water. The anthracite vein was not more than 18″-20″ high, and the mainways were never less than 3′. Little wonder that children were employed!

Stokes stated that only four people were allowed to descend the shaft at any one time, but admitted that new and stronger chains had been fitted to the winding engine following an accident in the previous year in which three people had been killed.[107]

Unfortunately it is not clear which was the principal shaft at Hean Castle at the period of Frank's visit, although it may have been an engine shaft situated on the hillside conveniently close to Coppet Hall beach. At all events, it seems that almost all the anthracite mined was won from the Lower Level vein, this being reached at different depths in the different pits on the estate: Kilback, 44 yards; Parsonage, 80 yards; New Mill, 60 yards; Engine Pit, Maryland, 60 yards; Phillips Pit, 54 yards; Thomas Pit, 54 yards; Hands Pit, 54 yards. No figure has been found for Cants but at one time the pit may have been entered by a slant as well as a shaft.[108] It seems clear that below ground here there was a bully or drift 150 yards long running towards the south-west. Cants, Maryland and New Mill are thought to have been active in the 1830's, but all seem to have been closed by 1850 except Cants, which by then may have been linked to Bonvilles Court. No date is known for the closure of the colliery at Coppet Hall, but again 1850 seems a likely date. On 1 December, 1851, Thomas Stokes agreed to lease the small pit at Betty Brown's Point, Coppet Hall, to Thomas Rees, giving him the right to work the Rock and Upper Fiddlers' Veins. Rees does not appear to have taken up the lease.[109]

Hean Castle (Woolsons and Speculation) The Stokes' family's Hean Castle colliery was almost certainly based on shafts close to the shore at Coppet Hall, but there is just a possibility that it included a pit on the estate at Woodside, Wiseman's Bridge, better known as Woolsons colliery. However there is no evidence for the existence of this shaft prior to the 1840's, and a local story that the pit was sunk primarily to obtain fireclay may well be true. Thomas Stokes certainly took an interest in all the minerals on his estate, and by the mid-1840's he was extracting coal and iron ore from levels or "patches" driven into the cliffs south of Woolsons colliery. At the same period he built a firebrick works on land adjacent to the colliery, and a steam powered corn mill at Coppet Hall. In 1845 the Saundersfoot Railway & Harbour Co. regarded

Stokes' enterprise with satisfaction because it offered the prospect of morc trade, but by 1847 there were misgivings. The directors learned then that Stokes was using company land beside the railway for depositing iron ore, fire clay and other materials obtained from the cliffs, and had also laid rails across the company's line to enable him to transport these materials. Just to add to their anxiety, the company's officers considered that cliff falls and damage to their property "were mainly, if not wholly, attributable to the levels driven into the cliffs by Mr. Thomas Stokes".[109] The matter was handed over to the lawyers, but without rapid results. Two years later it was alleged that Stokes was encroaching on S. R. & H. Co. land, and in June 1850 litigation was still unresolved.[110]

This dispute may have dampened Stokes' enthusiasm; by 1850 he was certainly less involved with the colliery. In 1849 it was reported that he had "let his colliery and iron mine to a party in London with a view to the formation of a new iron company".[111] If this happened, it was short lived, but the name "Woolsons" may date from this period. Certainly no record of the name has come to light before 1850 and there is ample evidence of it afterwards. On 1 December, 1851, Stokes granted a 60 year lease of part of Woolsons colliery to the local engineer, David Parsell. The lease included the use of a steam engine to pump water out of the pit, and of a remelting cupola with blowing machinery.[112] He was allowed to use the nearby Hill pit for ventilation, and unidentified pits in land held by Dinah Prout. In addition he was entitled to make free use of blacksmiths' equipment, and joint use of iron wagons with the tenants of the adjacent brickworks.[113] A further 60 year mining lease was granted by Stokes to Parsell on 1 January, 1852.[114] On the strength of these leases, on 5 April, 1852, Parsell enlisted the assistance of John Wall and Thomas Hill in working the colliery, these two men receiving half the profits and paying all expenses involved in sinking a pit to the Low vein from a point near Woolsons.[115] Some problem must have been encountered because in October, 1852, Parsell turned his lease over to Stokes.[116] There is no evidence of reletting, and by January, 1854, Stokes was negotiating the sale of the whole estate to his brother-in-law Edward Wilson,[117] who had leased Hean

Castle coal and iron mines for a brief period in 1844/45. As owner of the estate it seems that Wilson worked coal, at least until 1861. In the meantime, if a newspaper report of 1866 is to be believed, Thomas Stokes became secretary of the Pembroke & Tenby Railway.[118] He worked in this capacity from the early days of the railway until 1873. On his death in 1885 his body was interred at St. Issells.

It is not known when coal was last mined at Woolsons, but it appears to have been before 1873. In that year the pit was said to command access to a "very large tract of land under the sea".[119] In 1879 it was suggested that the colliery could be re-opened cheaply because a good part of the Lower Level vein had been unworked "when the colliery closed many years ago in consequence of dissension amongst the partners". The report went on to refer to the Speculation pit, in the wood north of St. Issells church. This shaft, 104 yards deep, was in process of being sunk by the same partners when the argument between them began. Work stopped at the same time even though the shaft was said to be close to the Kilgetty vein. Accordingly the 1879 report proposed that both shafts be put down to the Kilgetty vein, and a connection made between them. In this way "a fine colliery might be opened under the estate, admirably situated for the transmission of coals without detracting from the beauty of Hean Castle".[120] Quite possibly this had been the intention of the quarreling partners: significantly Woolsons' workings went inland towards the Speculation pit. One heading was driven a short distance due north from Woolsons, but seven others ran north-east, converging at an underground pond or sump beyond the road from Hean Castle to Kilgetty. In spite of the 1873 comment there is no sign that workings ever extended out under Saundersfoot Bay. However Woolsons seems to have been re-opened for trial workings in 1881,[121] but no new developments followed. Thereafter the shaft was fenced off and for many years covered only with timber. After the Second World War the site was given over to caravans, and the pit was properly capped. Even so, the shaft was broken into as recently as 1977 when work was being carried out on the caravan site drainage. The stonework was said to be excellent, although the pit itself was flooded.

Hean Castle Brickworks Although not part of the coal industry, the brickworks can be conveniently included here; it did in any case use black clay from the coal bearing strata as a raw material. Founded by Thomas Stokes in about 1845, the business developed quickly and in 1849 it was said that the quality of the firebricks was becoming well known.[122] By 1850 "a colony of people was engaged in turning out the best firebricks the county could produce";[123] the kilns were said to resemble "mosques". At this date the brickworks, like the adjacent colliery, may have been in the hands of lessees. After Stokes sold the Hean Castle estate the brickworks was certainly leased, the lessee in October, 1859, being William Walters.[124] In 1860, at the period of the accompanying plan, he appears to have been trading under the name of the Tenby Silica Co. By 1864, when Walters was associated with P. Cruikshank and W. W. Hayes the business was known as the Hean Castle Brick Co.[125] The bricks which originally had been marked "Stokes, Saundersfoot, Pembs", were now inscribed "Hean Castle, Tenby Pattern". A few examples of both types still exist.

In 1867 Davies & Roberts, contractors to the Pembroke & Tenby Railway, obtained firebricks from the "Woolston Firebrick works, Saundersfoot". Although the name was new, the brickworks were presumably Hean Castle.[126] All four kilns were in use in 1873,[127] and in 1879 they would seem to have been active. The works was then let at a fixed rent of £175 per annum, including the royalty in the fireclay. A report of 1879 described the firebrick works as "compact and complete, and well placed for conveyance and shipment, with covered drying floors and heating flues, four circular kilns, pig mill, steam engine etc". It went on to claim that "the bricks and slates have established a high character for durability in blast furnaces, coke ovens etc".[128] Unfortunately this reputation was not a guard against a time of recession, and by 1887 the brickworks was closed.[129] The exact date of closure is not known, but the buildings remained for a time, the drying shed and weigh house surviving until the 1930's. Although the area is now covered by caravans, foundations of the kilns came to light during excavations on the site in 1977/1978.

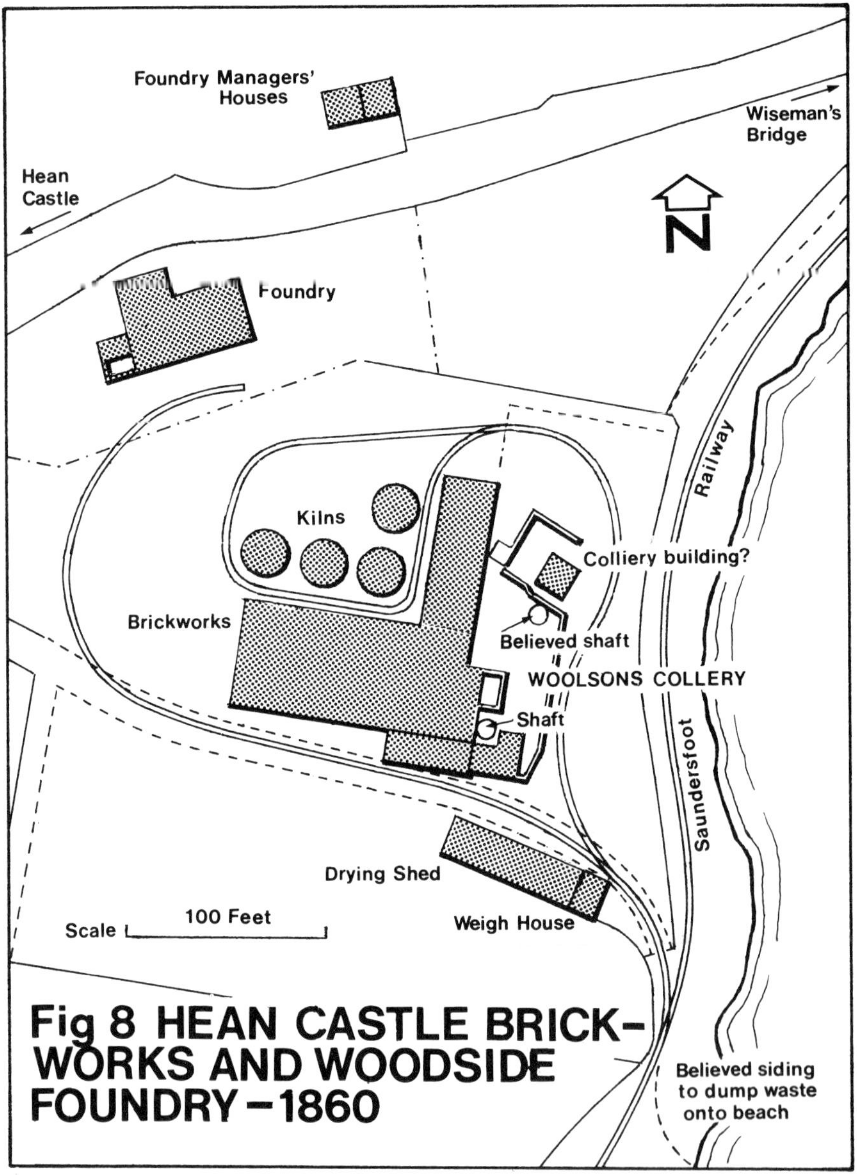

Fig 8 HEAN CASTLE BRICK-WORKS AND WOODSIDE FOUNDRY – 1860

Footnotes

[1]Vickerman: Dr. E. Frankland's report, 15.7.1853.
[2]P. I. C. 21.1.1857.
[3]Goggerdan, J. Woolstenhulme's report, 30.9.1778.
[4]Goggerdan.
[5]Goggerdan.
[6]L. L. 462/Goggerdan.
[7]L. L. 463 and 466.
[8]Goggerdan.
[9]Goggerdan.
[10]Goggerdan, J. Woolstenhulme's report, 30.9.1778.
[11]Goggerdan. Letter of E. Loveden to Lord Milford 20.7.1779; Reply of Lord Milford 23.7.1779.
[12]Wraxall, Memorials, v. 251, quoted by Sir L. Namier and J. Brooke in "The History of Parliament: The House of Commons, 1754-1790".
[13]Goggerdan.
[14]Goggerdan.
[15]Goggerdan.
[16]Goggerdan, Letter of H. Jeffreys, dated 29.3.1807.
[17]Goggerdan, Letter of E. Martin, dated 19.11.1807.
[18]Goggerdan, Letter of T. George, dated 11.3.1808.
[19]L. L. 469.
[20]Goggerdan, G. Daniel's Report to Sir Pryse Pryse, 1864.
[21]SRH, 1.6.1840.
[22]Thomas Evans, Inspector of Mines Report (1860) quoted by Edwards (1950).
[23]H. M. H. T. -.4.1862.
[24]Goggerdan, G. Daniel's Report to Sir Pryse Pryse, 1864.
[25]WW 13037. Also L. L. 476, 477, 479.
[26]Patent No. 3207/1871 granted to D. Parsell.
[27]Patent No. 1595/1873 granted to S. S. Williamson and D. Parsell. Patent 1596/1873 was also granted to both men.
[28]T. O. 30.4.1874.
[29]Pembroke & Tenby Railway Minutes 19.6.1873 and 23.12.1873 (P. R. O., Kew, Rail 559, Piece 2).
[30]T. O. 7.10.1869.
[31]The auction took place at de Rutzen Arms Hotel, Narberth on 29 and 31.8.1872.
[32]T. O. 30.4.1874 and Vickerman.
[33]L. L. 479.
[34]L. L. 481 and PC 4107.
[35]Vickerman: Report of Capt. H. Smith, c.1879.
[36]Vickerman: Valuation of C. F. Adams on Moreton and Trevayne estates 12.1.1882.
[37]NCB/2079 and R13541. N. C. B. records suggest that the Lower Level vein at Moreton was not worked for more than twelve years; presumably the workings were not very successful.
[38]H. M. H. T. 4.5.1887. Latterly the colliery agent was R. J. H. Fry and the manager, David Price.
[39]Vickerman: Memorandum of 26.11.1900. (See Appendix No. 2).
[40]Carew estate papers.
[41]Goggerdan: J. Woolstenhulme's report 30.9.1778.

[42] L. L. 459.
[43] EEW 6142-6148.
[44] L. L. 459, 462.
[45] L. L. 467.
[46] L. L. 467.
[47] L. L. 465, 466.
[48] NCB/2079 and R13541.
[49] L. L. 484. The Royal Commission on Land in Wales in 1894 heard Hilling allege that C. R. Vickerman had had his family evicted from their home at Vatson Moor simply on the grounds that Hilling's small pit was supposedly competing with Mr. Vickerman's collieries. The Commission seems to have taken this seriously, and also allegations upon other matters that had happened over twenty years earlier! (See note 24, Chapter One).
[50] PC 4083.
[51] L. L. 459, 460, 465.
[52] Plans of Whitland and Milford Haven Railway, 1863.
[53] NCB district plan.
[54] L. L. 488, 490.
[55] SRH 13.9.1833.
[56] L. L. 468. The manager at this period was David Morgan. Lewis Wilson seems to have had a long association with local industry. According to H. M. H. T. he was Harcourt Powell's agent in November 1862, and supported the proposed Whitland & Milford Haven Railway. Lewis Wilson of Langdon died on 27.7.1879 aged 68 and was buried at Begelly. William P. Wilson of Langdon is buried in an adjacent grave. He died in February, 1908, aged 53, as a result (it is thought) of an accident at Bonvilles Court colliery.
[57] Carew estate papers, No. 69 re lease 1.9.1839.
[58] PC 4100.
[59] GSWC p. 185.
[60] Plans of Whitland & Milford Haven Railway, 1863.
[61] T. O. 11.7.1867.
[62] PC 4106.
[63] L. L. 471, 474, 475.
[64] L. L. 473.
[65] Carew estate papers.
[66] L. L. 474, 478.
[67] PC 4107, L. L. 478, 481, 482.
[68] PC 4107, L. L. 482.
[69] Edwards (1950).
[70] PC 4107, L. L. 478, 481, 482.
[71] GWSC p. 159.
[72] Edwards (1950). Latterly the colliery was known locally as Hilling's Pit.
[73] Edwards (1950); NCB/5496.
[74] NCB/8762. E. Howells, who worked Wood Level in its last period (see Chapter Four) attempted to mine in this area in World War II, but the site was found to be too wet.
[75] NCB/10475; Department of Mines, List of Mines 1926-1930. Although the proprietor's name was once given to the writer as "Butler", these records indicate that he was W. C. Gunter. This pit appears to have been idle in 1928, but active in 1929 under the grand title of the Broadmoor Colliery Co. It is thought that at one period in the 1920's the pit was owned by J. Williams of Market Square, Narberth, and known as the Roxburgh pit.

[76]PC 4076.

[77]PC 4089.

[78]Howells p. 89.

[79]Vickerman. Myers' original lease was dated 29.9.1840, but colliery construction may not have taken place much before 1842, the year sometimes given for the establishment of the pit.

[80]SRH 3.6.1844.

[81]WW 12999.

[82]SRH 4.6.1849.

[83]WW 455. Myers appears to have obtained leases from J. M. Child in both 1840 and 1846. John Longbourne, who occupied the property known as Bonvilles Court under a lease of J. M. Child, was drowned in an accident at Egremont, Pembs, 5.8.1850 and buried at St. Issells.

[84]WW 13000.

[85]WW 13110.

[86]SRH Accounts for the period 1857-1861.

[87]PH 16.10.1863; Morris & Williams, p. 65. C. R. Vickerman himself, in a letter of 22.3.1878, said that he put up the ventilator not more than twelve years before. If this amounts to a discrepancy, perhaps the 1863 ventilator was a failure.

[88]Vickerman: Notes endorsed on a copy of Messrs. Lewis & Bidder's report.

[89]Vickerman: WW 13005.

[90]Vickerman: a letter of 10.3.1871 refers to J. Cadman being sacked some time before.

[91]H. M. H. T. 9.2.1887. The manager of Bonvilles Court up to 1884 was David John, and thereafter for a time Richard Williams.

[92]Vickerman: Memorandum of 26.11.1900 has annexed a list of leases then in existence.

[93]Pembroke County Guardian, 10.10.1913, reported the fire of 5.10.1913.

[94]The author is indebted to O. S. Gwyther and the late H. Prout for this description.

[95]John Christmas Samuel had worked in South African gold mines before moving to Wales. After the closure of Bonvilles Court he became manager of Hook Colliery, owned by Watts & Watts Co. Ltd. His successor with the Bonvilles Court Coal Co. Ltd was H. D. Connick.

[96]Pembrokeshire Telegraph 25.7.1929, 8.8.1929, 15.8.1929, 29.8.1929.

[97]L. L. 500.

[98]Pembrokeshire Telegraph 1.5.1930.

[99]L. L. 502.

[100]The new screens were built close to Bonvilles Court's standard gauge siding, and the sidings of the Saundersfoot Railway near the colliery pithead were realigned to serve it.

[101]Vickerman.

[102]6" O. S. maps, 1887 and 1906.

[103]Hean Castle estate map, 1764.

[104]Land tax returns, 1786, quoted by Stickings, p. 41.

[105]Hean Castle estate map, c 1845, in Hean Castle estate office.

[106]SRH 5.6.1843.

[107]Thomas Stokes evidence to the Childrens Employment Commission is clearly under the heading of Thomas Chapel colliery, but very curiously suggests that it was in St. Issells parish. On Page 471 of the Commission's report Franks gives a summary of Pembrokeshire collieries he had visited personally, and refers to Hean Castle colliery, Begelly parish, owned by Messrs. Stokes & Co., and to Thomas Chapel colliery, Amroth parish,

owned by Messrs. Morgan Hughes & Co. In fact Thomas Chapel colliery was in Begelly parish and Hean Castle colliery in St. Issells parish. Could Franks have been confused by the existence of Thomas' pit on the Hean Castle estate, close to Coppet Hall?

[108]NCB district map R13545.

[109]WW 13095.

[110]SRH letter from S. Giles to SRH proprietor 20.5.1847.

[111]SRH 3.6.1850.

[112]SRH 4.6.1849.

[113]WW 13094. An area near Phillips Pit on the Hean Castle estate is still sometimes known as Prouts pit, and this may be the area once worked by or for Dinah Prout.

[114]WW 13099.

[115]WW 13097.

[116]WW 13096.

[117]WW 13100.

[118]H. M. H. T. 5.9.1866. Thomas Stokes was secretary of the Pembroke & Tenby Railway from about 1861.

[119]Messrs. Lewis & Bidder's report, 1.2.1873.

[120]Vickerman. Report on Hean Castle minerals 24.3.1879.

[121]NCB abandonment records.

[122]SRH 4.6.1849.

[123]R. A. Mason "Guide to Tenby" 3rd edition, 34 (1854).

[124]SRH accounts. The Tenby Silica Co. may have been quite a separate concern, however. A man called Williamson is said to have had an interest in the brickworks after Thomas Stokes, and one or two bricks bearing the name Williamson are reputed to exist. No documentary evidence has been found to support this, although it may be significant that David Parsell who was involved at Woolsons colliery was also associated with Messrs. Williamson of Moreton colliery.

[125]South Wales, Tenby and Milford Haven Junction Railway plans, 1864.

[126]Llandinam papers, 6.

[127]Messrs. Lewis & Bidder's Report, 1.2.1873.

[128]Vickerman. Report on Hean Castle minerals 24.3.1879.

[129]6" O. S. map, 1887.

Amroth, c. 1910. View looking east

View south from Kilgetty Deer Park to Stepaside

(Kilgetty colliery is to the extreme left, behind the school. The Ironworks site is marked by the top chimney just visible in the trees on the left)

Stepaside, as seen from the bridge; Kilgetty Colliery to left

Kilgetty Colliery, view from Stepaside bridge

Stepaside and Kilgetty Colliery

Lower Level Colliery, view looking north

Locomotive ''Rosalind''

Saundersfoot Station

''Bulldog'' crosses road between Bonvilles Court and Saundersfoot Station

Saundersfoot Railway Bridge over main road at Begelly

''Bulldog'' and train pause on the railway near Begelly

''Bulldog'' in its original condition

Chapter Four

THE COAL INDUSTRY: COLLIERIES TO THE NORTH OF THE ERROXHILL FAULT

Generally speaking, the collieries north of the Erroxhill Fault obtained anthracite from the Lower Level and Kilgetty veins, the pits usually being sunk some distance to the south of the veins' outcrop to enable the coal to be worked to the rise. Mining in this part of the coalfield continued throughout the first half of the twentieth century, and marked the final phase in the story of industrial Saundersfoot.

Loveston Loveston village lies about six miles north west of Saundersfoot, and its pits were the most distant of those tending to look towards the coast. As at Broadmoor and Jeffreston, however, it is likely that some of the output was conveyed to the River Cleddau for shipment. The earliest reference to mining at Loveston is dated 1795, when William Llewhelin and John Boston were working a colliery under lease from the Leach family.[1] It is thought that the principal shafts at Loveston in the early nineteenth century were the Hillas Pit and the Breeches Pit; the former was later known as "Old Loveston colliery". The Breeches Pit was drained by an adit to the south, which also served a number of other small shafts.

A report of 1874 indicates that mining was at a standstill at that date. The report, by Hort Huxham, an engineer of Swansea, estimated a total of almost 2½ million tons of coal to be won from six veins at Loveston. He advocated sinking a new pit on the Lower Level vein considerably to the deep of the old Loveston colliery, but using that pit as an upcast shaft, and crosscutting to the Garland and other seams. He claimed that if ample engine power was placed here to pump and wind an output of 50 or 60 tons per day could be maintained for many years. Huxham also proposed trials on the Timber, Kilgetty and Tin Pits veins, but in view of the slump of the mid 1870's it must be doubtful if any action was

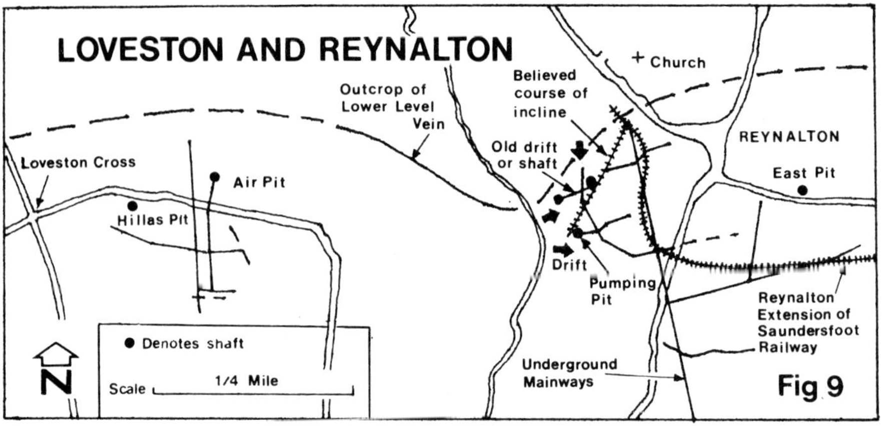
LOVESTON AND REYNALTON
Church
Believed course of incline
Outcrop of Lower Level Vein
REYNALTON
Loveston Cross
Air Pit
Old drift or shaft
East Pit
Hillas Pit
Drift
Pumping Pit
Reynalton Extension of Saundersfoot Railway
Denotes shaft
Scale
1/4 Mile
N
Underground Mainways

Fig 9

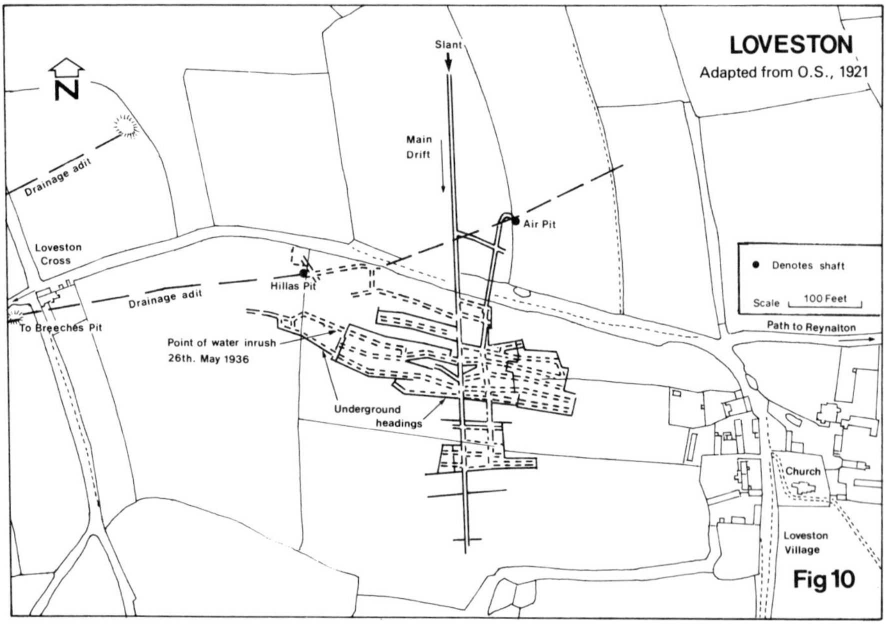
LOVESTON
Adapted from O.S., 1921
Slant
N
Main Drift
Drainage adit
Air Pit
Loveston Cross
Denotes shaft
Hillas Pit
Drainage adit
Scale
100 Feet
To Breeches Pit
Path to Reynalton
Point of water inrush 26th. May 1936
Underground headings
Church
Loveston Village

Fig 10

taken on his report.[2] By 1891 the only active pit in this area was at Yerbeston, a little further west.

Loveston's last period of operation began in 1932 when workings apparently based on the old Loveston colliery were reopened. By 1934 48 men were employed, and the number soon rose to 60.[3] Larger developments were hoped for, but on 26 May, 1936, disaster struck. Colliers broke through into old and uncharted workings, and seven men drowned as the pit flooded. Although production eventually restarted, it did not last long. The colliery closed in August 1937 with the loss of forty jobs. The output that year was just 1765 tons, little more than half the production in the peak year of 1934.[4]

Reynalton Reynalton was unusual in that it achieved greater prominence for its mining in the twentieth century than it did in the nineteenth. In this district the Lower Level vein was worked from a slant and a shaft in the valley south of the church, and also from a shaft on the north side of the road two hundred yards east of East Farm, where the depth is said to have been 70 yards.[5] Between this shaft and Hacket a north and south fault intervened with a downthrow to the west of 15 or 20 yards effectively restricting the extent of mining. It is thought that early in the twentieth century a slant was put down to the Kilgetty vein from a point north of Reynalton, near Carne mountain. The vein is reputed to have been only 3″ or 4″ thick, and so not surprisingly this working was not a success.[6]

In 1906 eleven men were employed in mining at Reynalton, but by 1914 the number was forty-eight.[7] The main reason for the rise was the formation of the New Reynalton Anthracite Coal Co. in 1913. This company began by opening a slant and shaft in the valley south-west of the village, but at the same period (1913-1915) negotiated with the Leach family for mining leases at Masterlands and East Williamston.[8] Although the correspondence was protracted, nothing seems to have come of this scheme —perhaps because of wartime conditions. In 1917 mining came to a halt even at Reynalton, and no traffic appears to have passed over the Reynalton extension of the Saundersfoot Railway in the twelve months up to December, 1918.[9] Work resumed in 1919 but was stopped again in 1921 by labour troubles. The colliery

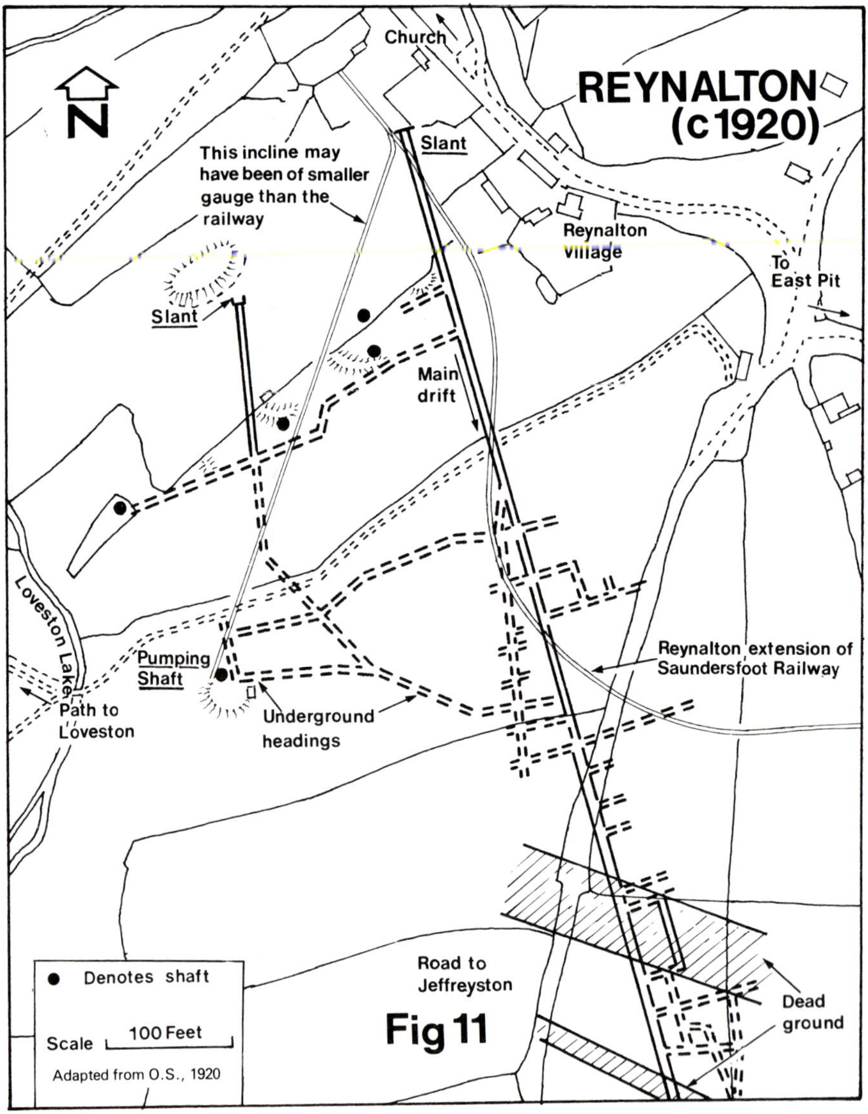

REYNALTON
(c1920)
N
Church
Slant
This incline may have been of smaller gauge than the railway
Reynalton Village
To East Pit
Slant
Main drift
Loveston Lake
Pumping Shaft
Path to Loveston
Underground headings
Reynalton extension of Saundersfoot Railway
Dead ground
Road to Jeffreyston
Denotes shaft
Scale 100 Feet
Adapted from O.S., 1920

Fig 11

never recovered, and by 1922 the decision had been taken to abandon Reynalton completely. A re-opening of the pit was rumoured in 1929 but failed to materialize.[10]

Thomas Chapel Early in the twentieth century C. H. R. Vickerman referred to the records of the Pembrokeshire Iron & Coal Co. to provide the following list of pits at Thomas Chapel:[11]

Old Wheel Pit sunk in 1784 to Lower Level vein, 53 yards
Lays Pit sunk in 1784 to Lower Level vein, 38 yards
Stump Park Pit sunk in 1790 to Lower Level vein, 54 yards
Hacket Pit sunk in 1794 to Lower Level vein, 58 yards
Air Pit sunk in 1799 to Lower Level vein, 36 yards
Lower New Hays Pit sunk in 1804 to Lower Level vein, 34 yards
Horse Wheel Pit sunk in 1836 to Kilgetty vein, 20 yards
Pond Landing Pit sunk in 1836 to Kilgetty vein, 32 yards
Pond Engine Pit sunk in 1840 to Kilgetty vein, 48 yards

The list is not complete, because quite apart from some crop-workings, it is known that there were other shafts in the vicinity of Hacket, Shipping and Broom. The name "Thomas Chapel colliery" was undoubtedly applied to different pits at different periods of time, and when the S. R. &. H. Co. was authorised in 1829 almost certainly related to the most northerly shafts at Thomas Chapel. By 1840 the name can be associated with the Horse Wheel Pit, the Pond Pit and the Engine Pit which was then being sunk. The S. R. & H. Co. papers for 1840 include a rather teasing reference to the Pond Pit "north of New Hayes near the old Engine House", and the company minutes mention that one pit was at work raising culm and a little coal, and another pit was ready to be worked. The pit in process of sinking had gone down 13 fathoms of an intended 27 fathoms.[12]

Most, if not all, the Thomas Chapel pits were located on Picton estate land. The principal lessees at this period were Morgan Hughes & Co., and their manager was said to be Lewis Wilson—possibly one and the same person as Lewis Wilson involved with the collieries at Broadmoor.[13] It was a time of troubles for Morgan Hughes. In January, 1842, he was described as "exceeding unlucky": the breaking of a shaft had delayed the construction of a driftway joining the old and new pits underground, and the old pit flooded.[14] The difficulties must have been considerable, because when R. H. Franks visited the district it seems Hughes had no employees at all. By June, 1845, the tone of the S. R. &. H. Co. proprietors was scarcely sympathetic. Morgan Hughes & Co. was defaulting on its debts, and the concern was said to be "sadly mismanaged". Even so, Hughes managed to subscribe to the T. S. & S. W. R. & P. Co. in the following year.

For several years Thomas Chapel colliery eked out a precarious existence, not in any way helped by the opening of Bonvilles Court, which, by virtue of proximity to Saundersfoot, had a considerable advantage in transport costs. In June, 1850, the general situation was still said to be unsatisfactory. In 1851, however, the Thomas Chapel pits were acquired by the Pembrokeshire Iron & Coal Co., doubtless as an additional source of fuel for the Kilgetty furnaces.[15] Some measure of output thereafter can be found in the fact that in March, 1852, 960 tons of coal or culm was produced. In February, 1854, the output was just 16 tons of coal and some 779 tons 5 cwt of culm. Perhaps the disproportionate amount of culm discouraged further production.[16] At all events, not long after it was decided to restart mining at Hacket and New Hayes, and once these pits were at work it appears that the more northerly Thomas Chapel colliery closed down.

Hacket and New Hayes Although part of the Thomas Chapel group of collieries, the New Hayes colliery of the 1850's was a new development on an old site. The scheme was an attempt to open up further workings in the Kilgetty vein by the sinking of a new shaft. The re-opening of Hacket colliery at the same period may be an indication that the two pits were linked underground, but whereas the new shaft at New Hayes went down 155 yards there is no evidence that Hacket was ever more than 58 yards deep. Hacket was back in production by July 1855, and in September, 1855, turned out 866 tons 18 cwt of coal and culm. However Hacket seems to have ceased production in June, 1856, and output thereafter is attributed to New Hayes.[17] If there was an underground link it may be surmised that whatever coal was won from either pit was wound to the surface at New Hayes, which was adjacent to the Saundersfoot Railway.

New Hayes was not a success. Although output was good in the

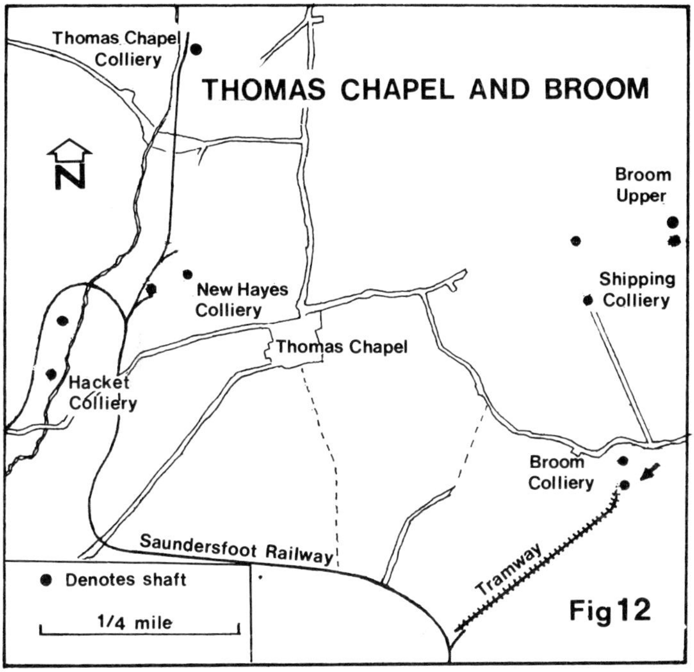
THOMAS CHAPEL AND BROOM
Thomas Chapel Colliery
Broom Upper
New Hayes Colliery
Shipping Colliery
Thomas Chapel
Hacket Colliery
Broom Colliery
Saundersfoot Railway
Tramway
Denotes shaft
1/4 mile
Fig 12

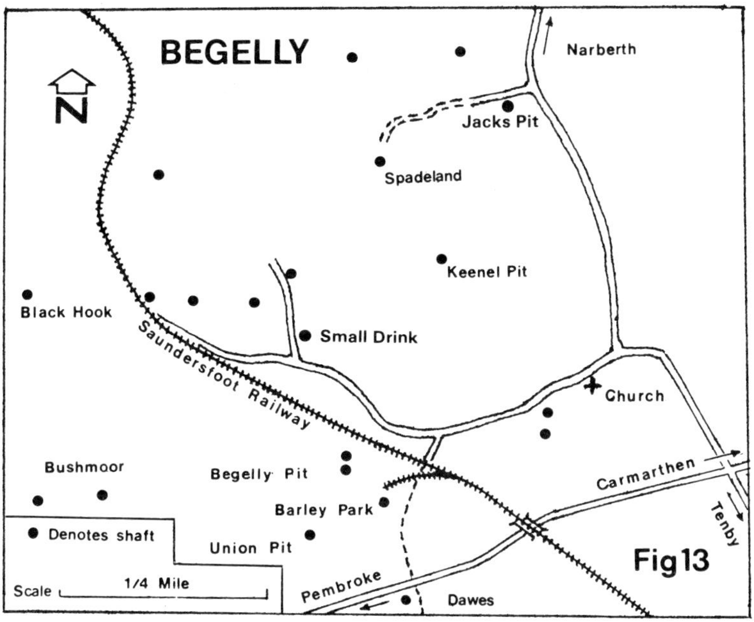
BEGELLY
Narberth
Jacks Pit
Spadeland
Keenel Pit
Black Hook
Small Drink
Saundersfoot Railway
Church
Bushmoor
Begelly Pit
Carmarthen
Barley Park
Tenby
Denotes shaft
Union Pit
Scale
1/4 Mile
Pembroke
Dawes
Fig 13

early months (over 1,238 tons of coal and culm were raised in October, 1856) serious difficulties were encountered and the colliery was abandoned in or by 1859. Writing to the landlord, Rev. J. H. A. Philipps of Picton Castle in November, 1859, C. R. Vickerman described what had happened. Having found the Kilgetty vein in excellent order, they had opened it out and worked it for several months before pausing to expend thousands on engineering means and appliances to work and land coal for many years. Scarcely had working resumed "than the vein fell away in all directions". Faults and water appeared instead, and the colliery had to be abandoned "with the loss of every shilling of outlay". Such a result, he added, "though rather more summary than usual is anything but an exception to the rule in Pembrokeshire collieries".[18]

In spite of the setback there were lingering hopes of a revival at New Hayes. One reference to the pit is dated 1867, although it must be very doubtful if there was any activity then.[19] As late as 1900 a steam engine was said to be in situ at New Hayes, but revival never came.[20] Today the site is completely cleared, although just across the valley at Hacket a ruined engine house still stands—over 120 years since coal was last raised!

Shipping and Broom Whilst New Hayes colliery was being redeveloped in the 1850's, efforts were made to find coal further east on Picton estate property near Thomas Chapel. By 1859 an experimental colliery was being sunk at Shipping, but the failure of New Hayes and the financial problems of the P. I. & C. Co. caused operations on this site to be suspended in May, 1859.[21] It may be assumed that the mining engineer Joseph Cadman, who was associated with C. R. Vickerman for some time, was disappointed by the decision, or at least felt he could do better. In July, 1861, he leased 500 acres at Thomas Chapel and Shipping with the reported intention of sinking a new pit at "Thomas House".[22] Two years later the local press noted that anthracite from Mr. Cadman's "Shipping Colliery" was being used by locomotives on the newly opened Pembroke & Tenby Railway.[23] Cadman seems to have enjoyed a closer relationship with the P. & T. R.'s contractors, Messrs. Davies & Roberts, than his employer C. R. Vickerman. Davies & Roberts appear to have been the lessees of the limestone quarries at Black Rock, Tenby, and in June, 1868, the local press reported that here "J. Cadman of Bonvilles Court" conducted the first local experiment with dynamite. It demonstrated the great superiority of dynamite over the "compound of vallaneous saltpetre" previously used, and Cadman was said to be hoping to set up an agency for its sale in the district.[24]

Shipping colliery, like Bonvilles Court, appears to have had two shafts situated about 400 yards apart. No. 1 pit (sometimes called Shipping Upper) was some distance to the north of the Begelly-Thomas Chapel road, and No. 2 pit (actually a slant) was just to the south of it, near the tiny hamlet of Broom. Both went down to the Lower Level vein, No. 2 being 52 yards deep.[25] It is thought that other workings north-east of Shipping and Broom were not connected with the colliery; they may have been in use at the same period, or earlier, but most were sunk to shallower veins of coal. In any event Shipping colliery did not enjoy a long life, and mining in the vicinity of No. 2 (or Broom) pit ended in 1875, and at Shipping Upper by 1881, on the expiration of Cadman's notice of termination of the lease.[26] Latterly Cadman had considerable difficulties with his landlord, C. R. Vickerman, who presumably held a head lease from the Picton estate. On one occasion in 1874 Vickerman told Cadman that his coal was "inferior and unfit for maltsters". Vickerman had other grounds for complaint too, because Joseph Cadman was often in arrears with his payments.[27] Finally, even after Cadman had left Shipping in 1881, Vickerman declared that he had no interest in purchasing any of the equipment because the colliery was in such an unsatisfactory state.[28]

The belief that good coal was still to be found survived the closure of the colliery. However, apart from a survey of the workings in 1919, there was no serious attempt to resuscitate the pit for over fifty years.[29] Then, following the closure of Bonvilles Court, Broom became the subject of a development scheme drawn up by Messrs. Wales and Lynch-Blosse, and Mr. Guy de Warren, who were well-known mining engineers. The plan envisaged working Picton estate minerals mostly lying to the south of existing workings.[30] Preparations on site began in November, 1932, and pumping operations began in the following spring. A new slant or drift was driven in a south-westerly direction with parallel headings to the west and east. The old No. 2 pit slant nearby appears to have been incorporated in the workings as an upcast; this pit was now said to be 60 yards deep. In the course of the works a 20″ seam of good quality anthracite was encountered, encouraging the proprietors to instal modern electrical machinery. The colliery duly re-opened in 1934, but the seam was soon found to be irregular and of very variable quality. The Bramblehill fault, which had often limited operations at Thomas Chapel collieries in the past, proved to be an obstacle in working westwards, and the whole pit was affected by water. In spite of a lot of expenditure the difficulties could not be removed. Even so, in 1936 Broom produced over 30,000 tons, and the output achieved in the two years following was almost as high.[31] By the end of 1938, however, the problems had become overwhelming, and the colliery closed down in January, 1939. Dismantling quickly followed, and within a few years even the waste tips had been largely cleared. Now apart from some disturbed ground and a small ruined building there is little to be seen on the site. Further down the hillside, however, there are remains of the incline specially constructed in the 1930's to take anthracite down to the Saundersfoot Railway. Although the sidings, like the machinery, have long gone to the scrap merchant, a few portions of rusty rail protrude from the colliery waste—a modest but adequate reminder of the activity once seen on the spot.

Begelly The story of mining at Begelly is intimately linked with the Child family who for many years lived at Begelly House. Early in the nineteenth century the Childs owned much of the property in and around Begelly village, although some land belonged to the Picton estate. The earliest records of Begelly colliery, dated 1776 and 1777, indicate that the pit was being worked on behalf of Lord Milford. In 1787 he was still in receipt of the profits.[32] The Begelly colliery belonging to the Child family may have been a separate concern: the Childs seem to have been running it soon after the turn of the century, Mrs. Sarah Child, and later her son James Mark Child being the most prominent characters. As noted in Chapter Six, these two were involved in the 1824 application for a licence to lay a tramroad down to the seashore. By 1830, however, the business was trading under the title of the "Tenby & Begelly Coal Co.", and J. M. Child almost certainly had a sizeable shareholding.[33]

The 1830's seem to have been a period of vigorous progress for Begelly colliery. About three-quarters of the output was shipped through the newly completed Saundersfoot harbour, making good use of two small vessels owned by the company itself—the "Peggy" and the "Mary Anne".[34] By 1836 the company had its own wharf at the harbour (the "Rock wharf") facilitating the growth of coastal trade. There was always a busy trade at the pit-head in coal for local use. In this time of prosperity, though, the coal company became involved in two arguments. The first was external, relating to the ownership of rails on a ¼ mile siding built to serve the pit at Barley Park. It was settled in February, 1840, when the S. R. & H. Co. sold the rails to the colliery.[35] The second was internal, involving differences of opinion between J. M. Child and Messrs. Pocock, who seem to have purchased their way into the business in 1838. By June, 1843, Child was reported to be ready to end their partnership: it must be very doubtful if he did, because he was undoubtedly the dominant figure in the later years of the colliery's career.[36] There is also evidence that Lewis Pocock was still involved as late as 1853.

R. H. Franks had some interesting things to say about Begelly at the time of his visit in 1841 or 1842. At that date the colliery had 60 h.p. of steam power, "winding and water engines". Underground trams were drawn distances varying from 10 yards to 200 yards. Few children under 10 were employed, the work requiring strength as the roads and roofs were soft "often closing works from superincumbent pressure". (sic!) Accordingly timber was an expensive item, and poisonous gases were also a problem. The thickness of the coal varied considerably, from 7′ to 15″: most of this is believed to have come from the Timber vein. As has been seen in Chapter 2, the colliery manager Richard Brough painted a somewhat cosy picture of the miners' life; J. M. Child was surely more candid in saying that women worked in the mine harder than the slaves in the West Indies!

In 1845 the manager was said to be sick, causing the Tenby &

Begelly Coal Co. to be "sadly in arrears" with its payments. There were also complaints that the company damaged parish roads by hauling timber over them instead of making use of the railway.[37] In the 1850's the proprietors were no more popular with the men than with the wider public: at this period Begelly is believed to have been the scene of industrial unrest. The company was still active in 1863, but by then it was in decline, and at least two of its earlier pits were out of use. One of these was Barley Park, which in 1845 had provided coal from the Rock and Low veins. The other was the Engine shaft (about 150 yards south of Begelly Rectory) which went down 106 yards to the Timber vein. This may have closed as early as 1852;[38] in 1853 Lewis Pocock was able to offer the P. I. & C. Co. a redundant 70 h.p. pumping engine.[39]

At least two pits were at work in 1863, namely the Union pit and Bushmoor. The Union pit (400 yards south-west of Begelly Rectory) reached the Timber vein at 120 yards. Although owned by J. M. Child, it was in the occupation of Griffith Thomas, very possibly a manager for Child or the Begelly company. Bushmoor was both owned and occupied by Child, and latterly had a shaft going down 74 yards to the Timber vein.[40] In 1845 the Rock and Low veins were worked at Bushmoor, the one being about 28" thick and the other no more than 6"-12" thick.[41] The Picton papers refer to an 1853 lease of the Bushmoor and Kingsmoor collieries by Richard, Lord Milford to Bridges Harvey. At first sight this seems contradictory, but there may well have been a second pit known as Bushmoor, located entirely on Picton estate land nearby.[42] Much of Kingsmoor, south of Begelly, was also part of the Picton estate, and although it had an active pit in 1830, the area was bisected by the Erroxhill fault and regarded as largely "dead ground" for coal mining.

A few other coal works at Begelly merit passing mention. In 1845 trial borings were taking place at Old Bush and Dawes, and it seems likely that both produced coal for a short period.[43] Black Hook colliery, further north, was older and more important. In 1806 Lord Milford granted a 21 year lease to T. Bowen and J. Brace,[44] and in 1829 it was of sufficient significance to win a reference in the plans of the S. R. & H. Co. It may even have had its own railway siding, but it seems to have been closed by 1840. Black Hook and Keenel pit were amongst several small pits drained by an adit running westwards to the Cresswell stream. The two shafts at Spadeland, north of Begelly, may have had drainage in the same direction, but whereas Black Hook was on the Timber vein the workings at Spadeland were in the Rock and Low veins.[45]

Exceptionally, for the area north of Erroxhill, there is no evidence of any attempt to work the deeper Lower Level and Kilgetty veins at Begelly. Instead mining came to an end at some date in the middle or late 1860's. In 1869 J. M. Child told the press that he had let the Begelly collieries to Mr. Page, civil engineer to the Board of Trade, but there is no sign that Page ever worked at Begelly.[46] Child himself died only a year or two later, and his son James Mark Phillips Child was never as deeply involved in local industry. In fact J. M. P. Child died in 1877, and his son Albert Theodore Hook-Child moved away to Brightlingsea, Essex.[47] In 1886, according to official records, the Timber vein at Begelly colliery was being worked again. The owner or agent was said to be Thomas Morgan of Begelly Lodge, and the manager John Powell.[48] No other reference has been found to the pit at this period, and it seems probable that whatever activity took place was short-lived. In the meantime A. T. Hook-Child retained property at Begelly and Bonvilles Court, but thereafter claimed rents and royalties as an absolute landlord. The Hook-Child interest continued right up to the 1930's, but by then the collieries at Begelly had all been cleared away.[49] Apart from some rough ground and a few small spoil tips there is very little to be seen at Begelly today.

Lower Level and Grove Records of the P. I. & C. Co. extant at the turn of the century enabled C. H. R. Vickerman to compile the following list of pits at Lower Level:[50]

Landing Pit sunk in 1797 to Lower Level vein, 12 yards
Landing Pit sunk in 1873 to Kilgetty vein, 106 yards
Engine Pit sunk in 1801 to Lower Level vein, 66 yards
Grove Pit sunk in 1856 to Lower Level vein, 132 yards
Grove Pit sunk in 1856 to Kilgetty vein, 210 yards

Leaving the references to Grove aside, it is clear that this list is

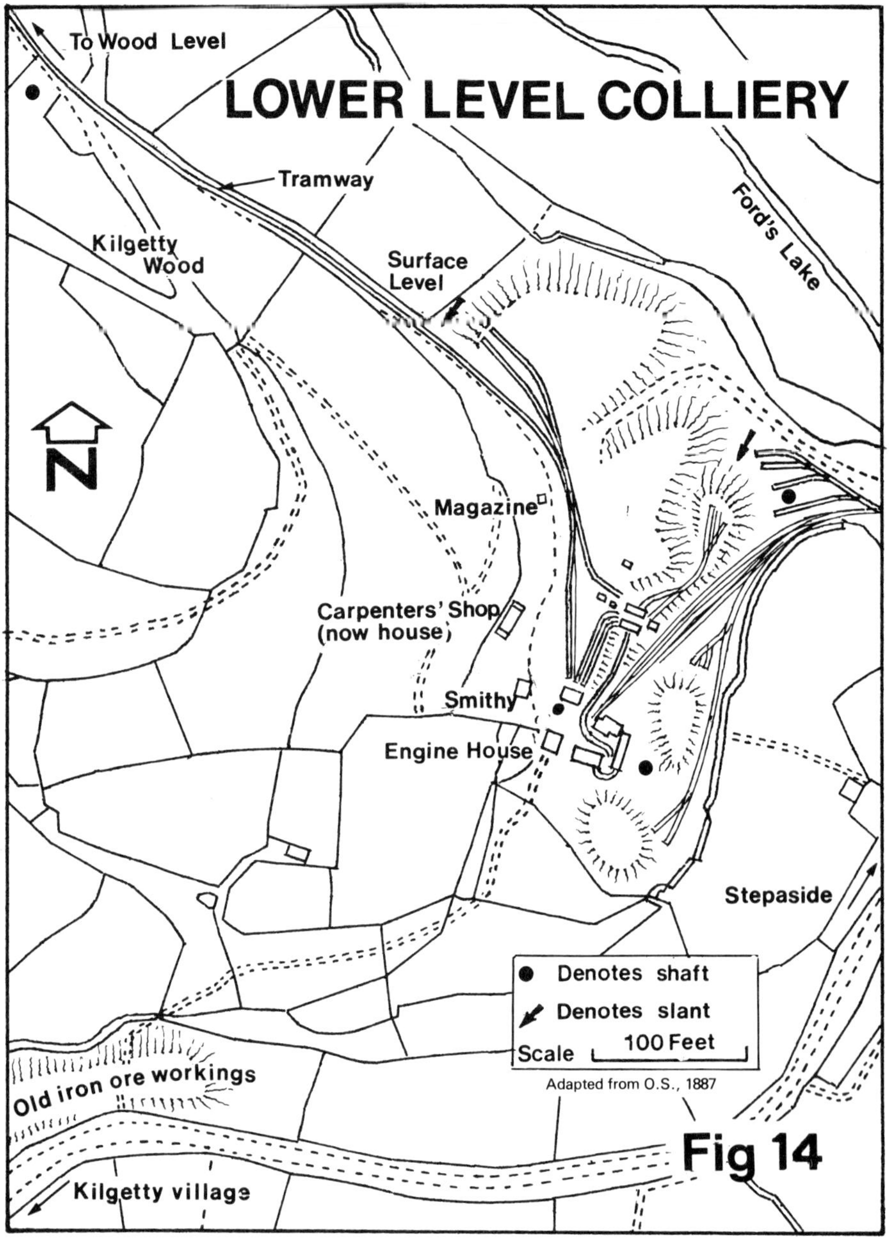

Adapted from O.S., 1887

Fig 14

not complete. First, there may well have been mining in this area prior to 1797, and although this cannot be proved there is evidence of a dispute as early as 1805 between Richard, Lord Milford and his tenants, J. Eynon, D. Williams and J. David over payments for coal extracted from land at Penrath, Little Kilvelgy and Cooksyeat.[51] Secondly, the Picton estate did not own all the land in the neighbourhood, and even the P. I. & C. Co. with its Picton estate associations, did not confine itself to Picton land. In the 1850's the company was working coal under both Picton land and Greville family land from Lower Level colliery, and also from the nearby Surface Level.[52] The name of Robert Fulke Greville, incidentally, is enduringly connected with the development of Milford Haven. He seems to have invested so heavily in that scheme, however, that the family property at Begelly and Kilgetty had to be sold in 1862.[53] At all events, the incompleteness of P. I. & C. Co. records may be partially explained by the presence of other landlords. Although it is known that in the early or mid nineteenth century, for example, there were workings at Lower Level known as Bows Pit and Mill Pit, no additional information is available. The Furnace Pit, probably of the same period, may have been an alternative name for one of the shafts already listed.[54] Further north, in Kilgetty Wood, there were several more drifts or shafts, the most notable of which was Wood Level. This pit will be described later in the chapter.

Little is known of Lower Level colliery prior to the formation of the P. I. & C. Co. In 1822 Richard, Lord Milford, leased the pit for forty years to J. H. Allen, H. Wilson, J. Griffiths and J. Mitchell.[55] How long the lease actually lasted is debatable: a Mr. Biddulph[56] seems to have had an interest in the pit by 1836, when the S. R. & H. Co. was extending its line north from Wiseman's Bridge. The story remains obscure until 1849, when the P. I. & C. Co. was clearly in control of Lower Level, sending much of its coal the short distance to the new ironworks. Lower Level was kept at work through the 1850's, whilst the Grove colliery was being sunk on the hillside overlooking the ironworks. The two pits were linked underground, and after Grove became operational Lower Level was made a downcast for the connected workings.

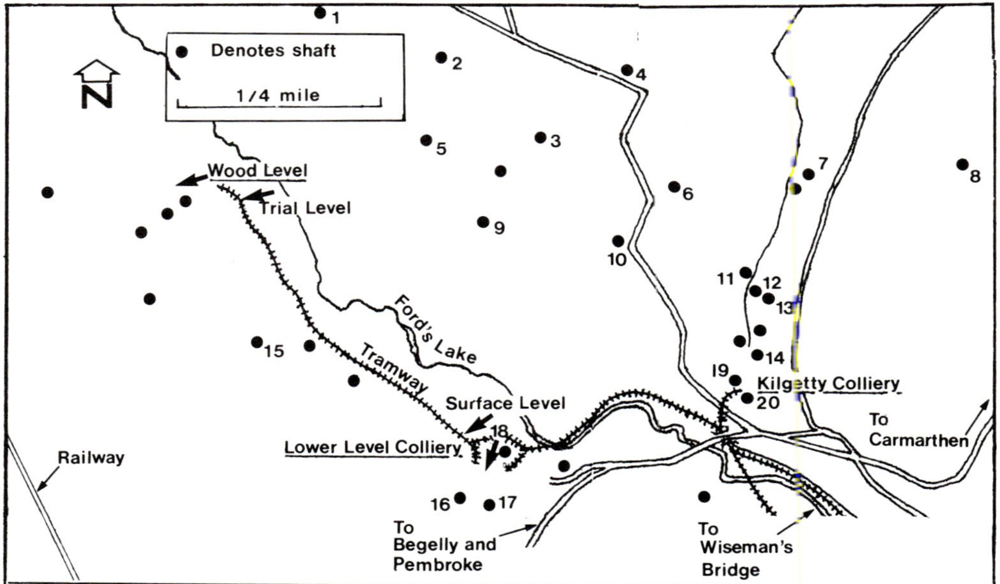

1	West Wood	8	Merrixton Top	15	Little Kilvelgy
2	West Park	9	Dingle	16	Landing Pit
3	Deer Park	10	White Gates	17	Engine Pit
4	Croft	11	Band of Hope	18	Bows Pit
5	Green Pit	12	Island	19	Wilsons Pit
6	Quarry Park	13	Whim	20	Engine Pit
7	Old Whimsey	14	Quarry		

Fig 15 LOWER LEVEL and KILGETTY

''Bulldog'', minus cab roof, believed to be at Broom

Loveston Colliery disaster—men leaving pit after rescue attempt

Edgar Howells, manager at Loveston and proprietor of Wood Level

Last years at Bonvilles Court: Iron stack being erected, 1927—
J. C. Samuel in foreground

Last years at Bonvilles Court: Howard Prout after last shift April, 1930

However as Grove was provided with a wide bratticed (divided) shaft, it may be surmised that both collieries had the means to operate independently.

The story of Grove colliery is the story of an imaginative and expensive venture which went badly wrong. The decision to sink the pit was taken in about 1852, no doubt as the fulfilment of Messrs. Fosters' 1845 recommendation that a new colliery be established at Kilgetty to serve the ironworks. The site was chosen not only to enable coal to be run directly downhill to the furnaces, but also to allow plenty of room for the disposal of waste. In spite of trial borings, geologically the location was unfortunate. The promoters had hoped to find the Kilgetty vein in a good condition, and then work it as far as possible. In the event a fault was encountered, and it took about five years (1853-1858) and over £30,000 to sink the huge shaft down 105 fathoms (210 yards) to the vein. It is believed that several people were killed or injured in the process. By September, 1859, the Kilgetty vein appeared to be in such a poor state that C. R. Vickerman doubted if they had reached it all, and so ordered the sinking of the shaft to resume in the expectation that it might have to go down another 25-35 fathoms to find a vein perhaps only 15″ thick. Understandably C. R. Vickerman revealed a degree of anxiety bordering on the desperate in his correspondence with the Rev. J. H. A. Philipps, the landlord.[57] At considerable length he described the precarious state of the enterprise, and appealed eloquently for reductions in the royalties on coal and culm. At first the landlord simply failed to respond, on the unhelpful grounds that the Kilgetty vein had yet to be proved. At a later date he may have offered some concession, but eventually Mr. Vickerman concluded that the 18″ of coal they had found at 105 fathoms was indeed the Kilgetty vein. The same shaft cut through the Lower Level vein at 43 yards where the coal was 21″ thick.[58]

When Grove colliery was complete there was something to show for the vast sums of money expended on it. In addition to the shaft, which C. R. Vickerman claimed in 1859 was certainly the largest in the county and perhaps "even in the Principality",[59] substantial pumping and winding engines were installed, together with an ample number of boilers to serve them. In

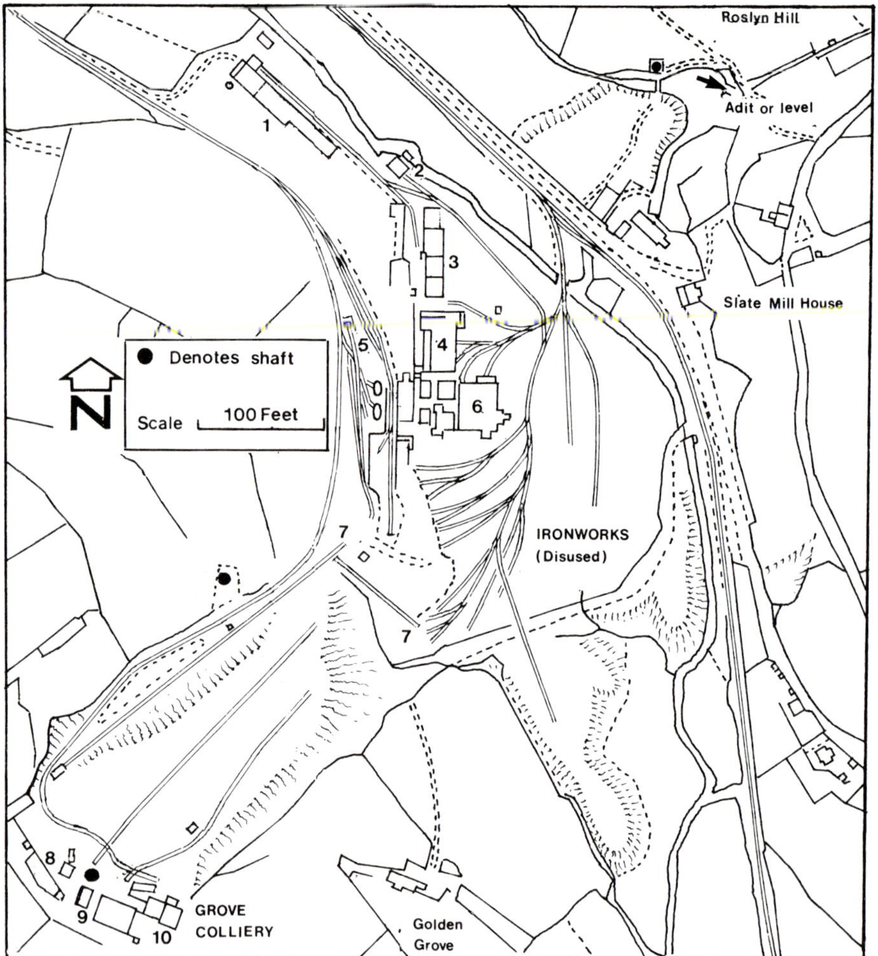

1 "The Platform" – workshops, smithy, sawmill
2 Locomotive shed
3 Boiler House? Coke Ovens?
4 Blast and Blowing Engine House
5 Furnace top and Kilns
6 Cast House and Refinery
7 Self-acting inclines
8 Reservoir
9 Engine House
10 Workshops, Stores, etc.

Fig 16 KILGETTY IRONWORKS AND GROVE COLLIERY – STEPASIDE (1887) Adapted from O.S., 1887

contrast Lower Level, although linked to the new pit, for a time seems to have been its poor relation, and may have seen only limited use. It was not until the formation of the Bonvilles Court Coal & Iron Co. Ltd. that the outlook improved for Lower Level. In the years 1873 and 1874 a shaft was sunk 107 yards down to reach the Kilgetty vein, and to mark the achievement Lower Level was renamed "Kilgetty colliery".[60] The previous Kilgetty colliery, centred on Wilsons, Quarry and Engine Pit, had closed a few years before, and the renaming may perhaps have been intended to maintain the name "Kilgetty" in the coal trade. Even so, the Kilgetty vein at Lower Level was not impressive. It was only 16" thick in the shaft, and as it was worked to the south mining was hampered by geological faulting.[61] Fortunately the problems were manageable, and in 1878 a brick arched underground roadway was built between Lower Level and Grove to enable them to operate effectively as one colliery. With the closure of the ironworks the importance of Grove diminished, and it is believed that the regular winding of coal there ended in 1884.[62] Thereafter the shaft was maintained primarily for pumping and ventilation, although it must be said that even in 1900 the colliery retained its equipment, including all facilities for winding.[63]

The underground workings of Lower Level and Grove extended south towards the Erroxhill fault and west towards Kilgetty village. Surviving plans suggest that roadways and headings were sometimes sinuous, and that underground haulage was extended by having to traverse tracts of "dead ground". At Lower Level colliery the main bully in the Kilgetty vein ran almost due south. Some of the workings were in ground which was not leased but owned freehold by C. R. Vickerman, and this may have been a factor encouraging production.[64] By the 1890's, however, anthracite was becoming more difficult to mine. Mr. Vickerman died in January, 1897, and although it is known that the Lower Level and Grove workings were very much in use later that same year, closure was not far away. Presumably workable coal reserves were almost exhausted, but regrettably research so far has not revealed exactly why or when the collieries closed. It is usually stated that Lower Level shut down in 1900, and this may well be correct, as regards most of the workings, even though there is a reference to the colliery being at work as late as November, 1900.[65] There is also evidence to suggest that for a couple more years efforts were made to mine coal at Lower Level, using the small Bows Pit and a nearby slant as well as the main shaft. This activity was probably on a very small scale, and ended in or by 1903.[66]

The Lower Level workings remained open at least until World War II, and one collier (and maybe others) descended the shaft in the 1930's astride a metal bar, or ''clatch iron''. He managed to get about eight hundred yards into the workings before meeting a rock fall. Now the shafts are capped, and the appearance of the site has been transformed. The carpenters shop has been converted into a private house, and the blacksmiths shop into a garage. Apart from the ruined shell of a winding house, only a small candle house and magazine are still to be found. In 1979 even the extensive spoil tips were being removed for road making and other purposes. More detailed study of the site was made difficult by dense foliage.[67] Much of the Grove colliery site is also overgrown, concealing the ruins of several buildings. The walls of the massive Cornish engine house are still easy to find, however, and nearby a large concrete slab covers the wide colliery shaft.

Wood Level In the eighteenth and nineteenth centuries numerous shafts or levels were created in Kilgetty Wood, with varying degrees of success. Surface Level, near Lower Level, prospered particularly in the early years of the ironworks: its workings extended north-west towards the property known as Penrath. Wood Level was probably the most important of these pits, and it appears to have been worked at three separate periods. The first occasion was in the late eighteenth or early nineteenth century, when colliers working the pillar and stall method left very large columns of coal to support the roof. It is not known exactly when this period came to an end.

Much later in the nineteenth century, when economic or technical difficulties had overtaken many other pits in the district, Wood Level was re-opened. On this occasion most of the columns of coal were removed, and the roof was shored up with timber. By now the colliery included two shafts, one of which was approached from within the main level. The workings are believed to have been about 100 yards deep, extending west and north-west from the pit mouth. This second period may not have lasted much more than a decade, and evidently came to an end in 1890.[68] In this short era Wood Level was provided with a small tramway through Kilgetty Wood to connect it with Lower Level colliery and also the Saundersfoot Railway. This tramway was probably operated by horses, and may well have served one or two other workings besides Wood Level.[69] The names Kilgetty Wood Pit and Wood Pit were associated with these workings at this time.

Even by 1890 the anthracite was not exhausted, but it was not until World War II that a local businessman, Edgar Howells, obtained rights to re-open the pit. Wood Level came back into production in 1944, and had an output of 4,000 or 5,000 tons per annum for several years.[70] This was all conveyed away by road. On the nationalisation of the coal industry in 1947 the pit was licensed as a private "small mine", as very few people were employed. Within a year or so it became clear that coal was increasingly scarce, and production appears to have ended in or by 1949. Mr. Howells maintained the pit for a short time whilst weighing up the future possibilities for mining at this and other sites. No developments ensued, however. A surviving plan of Wood Level is dated 30th June, 1950, and this may be taken as the time of final closure.[71] Edgar Howells himself probably hoped to continue this type of mining activity, but unfortunately died not long afterwards. However Wood Level under his direction secured a small claim to fame as the last active pit in the Saundersfoot area. Today the site of the pit is largely concealed by soil and vegetation, although the route of the small tramway is easy to follow as a forest track running down through the wood to Lower Level.

Kilgetty Early in the twentieth century C. R. Vickerman consulted P. I. & C. Co. records to obtain the following lists of pits at Kilgetty:[72]

Old Engine Pit sunk in 1775 to the Kilgetty vein, 6 yards
Croft Pit sunk in — to the Kilgetty vein, 76 yards
West Park sunk in 1780 to the Kilgetty vein, 84 yards
Quarry Park sunk in 1780 to the Kilgetty vein, 80 yards

Deer Park sunk in 1781 to the Kilgetty vein, 80 yards
White Gates sunk in 1787 to the Kilgetty vein, 84 yards
Engine Pit sunk in 1789 to the Kilgetty vein, 56 yards
Amroth Hill Pit sunk in 1798 to the Kilgetty vein, 84 yards
Bedlam Pit sunk in 1805 to the Kilgetty vein, 20 yards
William's Pit sunk in 1806 to the Kilgetty vein, 64 yards
Thomas Hill Pit sunk in 1810 to the Kilgetty vein, 94 yards
Pumping Pit sunk in 1840 to the Kilgetty vein, 99 yards
Wilson's Pit sunk in 1842 to the Kilgetty vein, 88 yards.

As in the case of both Thomas Chapel and Lower Level, Mr. Vickerman's list was far from complete. He must have known of the existence of other pits, but probably regarded them either as so insignificant or as so long closed as not to be worthy of mention. A curious feature of the list is the inclusion of four collieries (Amroth Hill, Bedlam, Williams and Thomas Hill) sited not at Kilgetty but over the hill to the east, near Amroth. Presumably they were included on the grounds of being either owned or managed by the P. I & C. Co. enabling them to be referred to as part of the Kilgetty collieries.[73]

Most of the early pits listed were located in the parkland attached to Kilgetty House, a mansion which once stood on the hillside about half a mile north of the main Carmarthen-Pembroke road. The writer believes that in the eighteenth century the name "Kilgetty" (perhaps a corruption of the less wieldy "Kilvelgy") applied primarily to this estate and its immediate neighbourhood. The hillside to the east was known as Merrixton, and the valley below as Deepcombe, Kilgetty or Merrixton Bottom. The name "Stepaside" appears to have been little used or very localised before the twentieth century, by which time "Kilgetty" had become associated with the settlement developing around the railway station on the P. & T. line. Kilgetty park is believed to have been the site of other shafts, including West Wood, Dingle, Green, Hill and Lambs Park. In some cases the exact location of the shaft is not known. The valley between Kilgetty and Merrixton, north of the main road, was the site of other collieries. Engine Pit, Pumping Pit and Wilson's Pit were in the valley, as were Quarry and Band of Hope, and older pits known as Island, Whim, Old Whimsey and Main Level. There was only one shaft of any note on the hillside just to the east named, appropriately enough, Merrixton Top.[74]

Repeated references to particular pits may incline the reader to forget that coal mining usually began with small scale surface workings where the veins outcropped. Both the Kilgetty vein and the Lower Level vein were worked extensively in this fashion near Kilgetty, but it is clear from Mr. Vickerman's list and other sources that more advanced mining was taking place in the second half of the eighteenth century. Charles Poyer Callen was a notable coal proprietor at this period, and three surviving documents deal with shipments made by him from "Deepcombe Bottom, the property of Lord Milford" in the years 1788 and 1789.[75] Callen, who appears to have been related to the well known Pembrokeshire family of Poyer, then resident at The Grove, Narberth, was involved in coal mining for many years. In 1811 and 1817 he signed leases with Richard, Lord Milford in respect of a colliery at Kilgetty.[76] In 1818, however, he surrendered lands previously leased for forty years, giving up possession of a water engine, workshops and stores at Kilgetty Bottom.[77]

At this time Callen was not the only coal proprietor in the neighbourhood. In October, 1801, Richard, Lord Milford granted two mining leases to W. Raymond and others. One dealt with land near Wiseman's Bridge,[78] in the parish of Amroth, and the other (for 31 years) related to land at Poor Shipping on "Lord Milford's Kilgetty estate". Poor Shipping was some distance down the valley on the Amroth side, and both these leases may have related to the small pit known as White Leys, near Wiseman's Bridge. Nearer to Kilgetty proper, there was one early reference to Stepaside in a 1791 lease, which lasted just one year, made between Richard, Lord Milford and Daniel Lloyd.[79] The Lloyd family of Laques, Llanstephan, Carmarthenshire owned a considerable amount of land between Kilgetty, Wiseman's Bridge and Amroth.

Very little is known of the Kilgetty collieries between the era of Charles Callen and the opening of the Saundersfoot Railway, although some indication of their development is provided by a plan in the records of the Neath Abbey Ironworks which relates to a 14″ beam engine for "Lord Milford's Kilgetty colliery".[80] The plan is dated 27 December, 1811, and is the earliest positive reference to a steam engine in the Saundersfoot district. It seems likely that at this period the main colliery was based on the Engine, Quarry and Band of Hope pits.

According to the records of the S. R. & H. Co., the principal agent of the Kilgetty Co. in 1838 was one Robert Kirke. In the following year new works at the colliery were being supervised by Thomas Dyson, engineer of the Bedford Level.[81] Presumably these works included the sinking of the Pumping Pit of 1840 and Wilson's Pit of 1842, and they may also have involved the opening of a pit further down the valley at the slate mill. Some caution may be necessary on this score, though, as there appear to have been two companies working at Kilgetty at this period. R. H. Franks, referring to the year 1841, mentiones a company called Heptonstall & Co. A little later S. R. & H. Co. papers refer to Wilson & Co., but in 1844 also mention an old Kilgetty company and a new Kilgetty company. The landowner was Sir Richard Philipps, who had succeeded to the Picton estate, and it is evident that by the time the Pembrokeshire Iron & Coal Co. had been set up in 1846 the Kilgetty Co. was under its control.[82] The Wilson of Wilson & Co. (and, no doubt, Wilson's Pit) cannot be positively identified, but it may have been Edward Wilson, brother-in-law of Thomas Stokes of Hean Castle, and later owner of Hean Castle estate between 1854 and 1863.[83]

Whatever the truth about the proprietors, Kilgetty colliery prospered at this period. In January, 1842, the Kilgetty Co. was said to be "in full work", having 1,200 tons of coal in stock, and raising several hundred tons each month from their two pits.[84] This was a creditable achievement in a colliery working a coal vein never more than 24" thick, and often much less. According to Franks, the colliery had a depth of 45 fathoms, with three engines "power together 56 horses, water power 75 horses". Richard Hare, the then agent, declared that the mine was well ventilated, and made the interesting assertion that they had ten pits with a circulation of air from one to another. The extensive nature of these collieries is also evident from their system of drainage. Several pits in Kilgetty park were linked by an adit which ran down from Croft Pit to the valley below. A more substantial adit,

Last years at Saundersfoot: ''Rosalind'' shunts at the harbour

Last years at Saundersfoot: Tram being tipped on coal shoot, 1937

Last years at Saundersfoot: Loading coal into sailing vessel

Last years at Saundersfoot: Coastal Steamer by coal shoots

Bonvilles Court Colliery partially dismantled, 1939

Bonvilles Court Incline, 1939

Coal Trams at Bonvilles Court, 1939

Dismantled railway between Bonvilles Court and Saundersfoot Station, 1939

Saundersfoot Harbour, 1963

Saundersfoot Railway bridge over road at Reynalton, 1963

2—6—2T No. 4107 at Saundersfoot Station, 1963

D.M.U. for Pembroke Dock enters Kilgetty, 1966

The Cast House, Kilgetty Ironworks, 1978

referred to in Chapter Two, is believed to have connected with most of the pits in the valley, running from a point under Merrixton due east to Thomas Hill before turning south towards the shore at Amroth.

The opening of the ironworks gave the Kilgetty collieries a boost. Anticipating increasing output, the P. I. & C. Co. sought estimates in August, 1849, for the cost of completing a 24″ steam engine, and for continuing to work the Slate Mill pit until the Lower Level vein was located. This must have been unsuccessful because Slate Mill pit (comprising a shaft and a level) appears to have closed in 1850.[85] In the same year, however, a new bully or underground engine was ordered for Kilgetty colliery, and during the revival of iron-making in 1852 and 1853 coal was raised at Quarry and Wilson's Pit.[86] For reasons which are not clear, production at Wilson's Pit seems to have ended in or soon after February, 1855, and by May, 1857, it was noted that the Wilson's pumping engine was being converted for use as a second blast engine at the ironworks.[87]

The story of Kilgetty colliery in the 1860's is obscure, but it is believed that coal mining gradually declined, all the smaller and more scattered pits falling into disuse. The main workings are thought to have closed in or by 1870, the name "Kilgetty" being transferred for a time to the nearby Lower Level colliery. Lewis & Bidder (1873) refer to the old Kilgetty colliery as no longer being at work, even though the Kilgetty Vein "in fine order and condition has been proved by a slant working to the dip of 330 yards, with a large range to the east and to the west". Two spare engines were then lying at the colliery, one being a Cornish pumping engine with a 40″ cylinder, and the other a horizontal engine with a 20″ cylinder and a 6ft. stroke.[88]

It appears that Wilson's Pit was briefly reopened in 1885, but possibly only for a survey of the workings. It was then described as Kilgetty (Wilson's Pit).[89] What happened at this date may never be known, but there is clear evidence that later, on the demise of Lower Level colliery, C. H. R. Vickerman wanted to redevelop Kilgetty colliery using machinery transferred to it from Lower Level. At this time, in 1900, the colliery was said to comprise Engine Pit, with a shaft 49 fathoms deep and 10ft. in

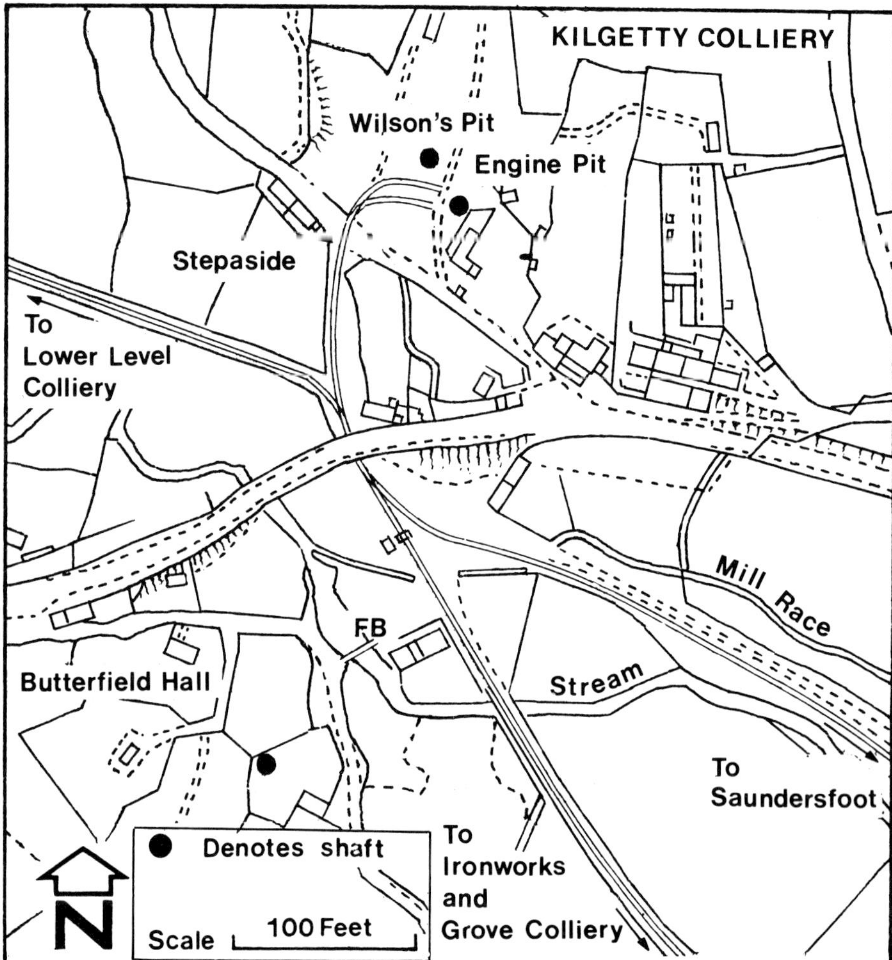

Fig 17 KILGETTY COLLIERY (1887)

Adapted from O.S., 1887

diameter and Quarry Pit, 42 fathoms deep and 8ft in diameter.[90] Although not specifically mentioned, Wilson's Pit must also have been accessible. There can be no doubt that there was more activity at Kilgetty in this period. Records show that the "Kilgetty Adit Level colliery" employed 45 men in 1902, 24 men in 1906 and 30 men in 1910. Unfortunately it is not stated which pit was the Adit Level colliery (most were drained by adit anyway) but this description seems to imply the use of a waterwheel on an adit for pumping purposes. In the twentieth century and perhaps earlier Kilgetty Colliery was sometimes referred to as "the Wheel Pit".[91] The available mining plans suggest that Wilson's, Engine and Quarry Pits were open at the end of 1902 but abandoned again during 1903. At the same time the small Mill pit (very possibly one and the same as the earlier Slate Mill pit) was opened 150 yards east of the ironworks. The Lower Level vein was found in an unworkable condition at 17½ yards and work was halted in 1903.[92] At the time of writing research has produced tantalising but insufficient information for any more detailed account.

In spite of the misfortune of the industry, coal mining at Kilgetty was not finished. A sizeable tract of unworked coal remained to the south and south-east of Wilson's Pit, and this caught the attention of Jasper G. Daniel, the mineral surveyor called in to advise on the prospects for the industry in the aftermath of the Bonvilles Court closure. In his report of April, 1930, Mr. Daniel noted that the company already had a winding engine and boiler installed at Wilson's Pit, with a stocking room and screening walls in position.[93] The directors were duly encouraged, but pumping operations at Kilgetty did not begin until 1934 or 1935, and then took much longer than expected. To assist in the task two colliers were asked to clear under Merrixton-Amroth adit. They entered it through Kilgetty colliery workings under Merrixton, and with some difficulty eventually negotiated their way right through to Amroth. After they had broken through, water gushed out under pressure for over twenty-four hours. The last obstruction of gravel actually burst out of the adit onto the beach whilst the two colliers were eating their lunch nearby! It must have been a satisfying end to a remarkable feat.[94]

By the 1930's the village in the valley was always known as

Stepaside, and so the Kilgetty colliery at this period was frequently referred to as the Stepaside colliery. The alternative name made no difference to its fortune, or lack of it. After reopening in 1937 under the auspices of the Kilgetty Anthracite Colliery Co.[95] it enjoyed no more than two years of steady production, the output rising dramatically from about 2,000 tons in 1937 to approximately 15,000 tons in the following year.[96] Although the anthracite found plenty of buyers, and much of it was dispatched to France, serious faulting was soon encountered in the Kilgetty vein. It was widely believed at the time that a fresh injection of capital would have overcome the difficulties, but the owners did not have the capital to hand. Appeals for government aid fell on deaf ears, and in January, 1939, the men were given notice. Arthur Horner, President of the South Wales Federation of Miners, declared that it was ''a scandal that in a case like this, where the company is anxious to work, and where there are prospects of remunerative results . . . that there should be no government assistance forthcoming''.[97] In February, 1939, the colliery closed, and 262 men were made redundant. A couple of months later Robert Williams, the colliery manager, put an end to rumours of reopening by leaving the district.[98] Most of the machinery, was dismantled within a year, and the site was allowed to decay. After World War II the shafts were properly capped, and in May, 1972, the main chimney stack was demolished. Apart from a winding house and one or two other buildings, there is little to be seen on the spot today. Almost all trace of the small pits in the valley north of Stepaside has been obliterated by tipping, although a few clues may remain buried in the dense undergrowth.

Amroth As may be apparent already, the history of coal mining at Amroth overlaps to some extent with the story of the industry at Kilgetty. As at Kilgetty, there are many uncertainties, but unlike Kilgetty it is not possible to say that one colliery was obviously outstanding. The four Amroth area pits recorded in the P. I. & C. Co. papers with the Kilgetty collieries were not large, and it is likely that only Thomas Hill pit (near Killanow) was still active in the second half of the nineteenth century. No closure date is known, but it does not appear to have outlived the P. I. & C. Co. itself.[99]

Thomas Hill, Bedlam and Williams' pits were all sited on or near the long adit running underground from Merrixton to Amroth. Three other pits on the same adit can be identified, one of which, Coombes colliery, was controlled at one time by the Kilgetty company, and probably later by the P. I. & C. Co. Sadly nothing is known of the other two pits, namely Skerry Back and Prouts' pits. Coombes, sometimes called "Old Coombes", was undoubtedly active in 1845, and it may also have been at work in 1853 when new colliery leases were granted.[100] It is thought to have closed soon after. However this pit seems to have had some history. The earliest record of a colliery near Amroth village (rather than simply within the parish) is dated 1793, and mentions pits called Hillsend and Coom Mills.[101] This may well be a reference to what was later called Coombes. The same may be said of a reference in 1824, when one Daniel David surrendered (at a price!) his lease of coal mines at Killanow, Little Merrixton and Coomshill.[102] It may be as well to add, though, that Coombes colliery was sunk in the valley only about half-a-mile inland from the sea, and worked coal in the Kilgetty vein. An interesting feature of the colliery was the provision of power from a water wheel on the drainage adit down from Merrixton. When the adit was reopened in the 1930's for the benefit of Kilgetty colliery, the colliers involved were surprised to discover two or three clay "nests" near the entrance to headings in the Coombe's workings. Each nest contained several clay balls—a reminder of the method of counting trams filled with coal in the days before the ordinary coal miner could count for himself![103]

A few other pits may be mentioned briefly. There is evidence of mining at Summerhill as early as 1797, when David Saer took a 21 year lease of coal and culm mines from Hugh Barlow of Lawrenny. Work on this site may have ended when Saer had the misfortune to go bankrupt in 1826, although this is far from certain.[104] Cwms Level will be mentioned in the next chapter as a source of iron ore, but appears to have produced some coal. Castlepark colliery was on the opposite side of the valley, and went down to the Kilgetty vein at 80 yards. This is believed to

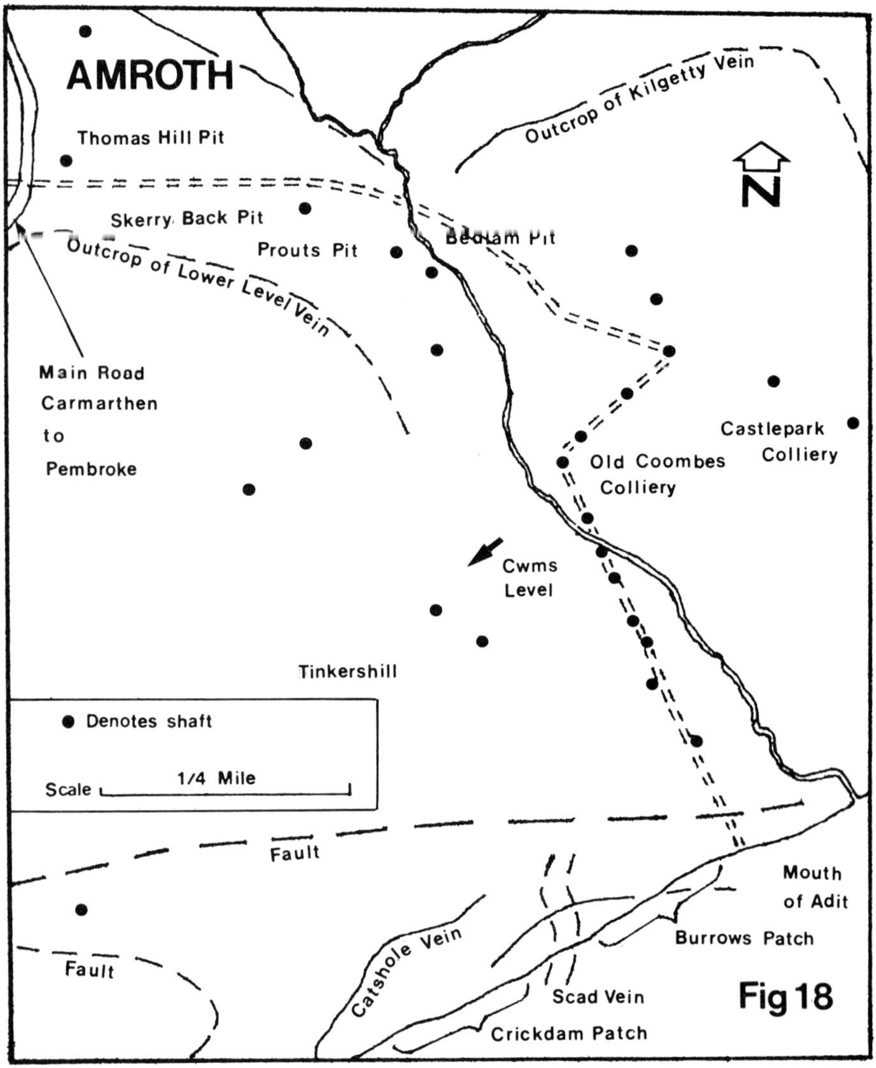

AMROTH
Thomas Hill Pit
Skerry Back Pit
Prouts Pit
Bedlam Pit
Outcrop of Kilgetty Vein
N
Outcrop of Lower Level Vein
Main Road
Carmarthen
to
Pembroke
Castlepark
Colliery
Old Coombes
Colliery
Cwms
Level
Tinkershill
Denotes shaft
Scale
1/4 Mile
Fault
Mouth
of Adit
Burrows Patch
Catshole Vein
Fault
Scad Vein
Crickdam Patch

Fig 18

have been a landing shaft linked underground with the Castlegate pit, a pumping shaft situated near the beach and the main entrance to Amroth Castle.[105] It is thought that the Biddulph or Biddolph family who were once resident at the Castle had an interest in these two pits, but unfortunately there is no evidence to hand. The records of the Neath Abbey Ironworks include an 1822 drawing depicting parts for an 18″ engine belonging to F. T. Biddolph, and it seems that the engine house was situated at Amroth.[106] The Castlepark and Castlegate collieries probably closed in the 1860's and today there is little to be seen of either. Indeed, apart from a few small spoil tips there is very little trace of coal mining ever having taken place around Amroth.

Footnotes

[1] L. L. 461.

[2] T. O. Report dated 1.9.1874.

[3] Edwards (1950). The manager at Loveston at this period was Edgar Howells, later associated with Wood Level. Lord Merthyr and a French associate were amongst the prime movers of the undertaking, which traded as the Loveston Colliery Co. Department of Mines, List of Mines 1934.

[4] Edwards (1950). The 1934 output was 3,420 tons.

[5] NCB district map.

[6] G. S. W. C. p. 180.

[7] Edwards (1950).

[8] L. L. 490-494.

[9] S. R. H. 3.6.1918 and 5.6.1922. Edwards (1963) p. 20 makes the point that rail rates operated against the colliery, 6d being charged on each ton of coal conveyed to Saundersfoot station, and a further shilling for transport from the station down the incline to the harbour. The peak year at Reynalton was 1919 when 11,215 tons were produced.

[10] Pembrokeshire Telegraph 22.8.1929.

[11] G. S. W. C. p. 177.

[12] S. R. H. 7.2.1840.

[13] S. R. H. 1.6.1840. R. H. Franks (1842) stated that Thomas Stokes & Co. were proprietors at Thomas Chapel, but as indicated in Note [107] to Chapter 3, details in Frank's own report suggest that this was incorrect.

[14] S. R. H. 11.1.1842.

[15] PC 4116 (P. I. C. Ledger).

[16] PC 4116. The purchase is recorded in the accounts of the P. I. & C. Co. for the period 1.4.1852 to 31.3.1853.

[17] PC 4116.

[18] PC 3830-3832.

[19] NCB abandonment plans.

[20] Vickerman. Memorandum of 26.11.1900.

[21] P. I. C. 31.5.1859.

[22]P. I. C. 23.7.1861. Joseph Cadman's brother, James Cadman, may have been involved in Shipping colliery at this period. James Cadman (b. 1812) was manager of Pentwyn Ironworks, near Varteg, Mon., in 1844, and in or about 1845 became manager of Tondu Ironworks on behalf of Sir Robert Price, M.P., principal proprietor of the Glamorgan Iron & Coal Co. In the 1850's James Cadman became manager of Maesteg Ironworks before going to Russia as an ironmaster. Joseph Cadman (b. 1813) did not travel so widely. In 1851, before moving to Pembrokeshire, he was manager of Cefncwse Ironworks, Cefncribwr, near Tondu, for Messrs. Malins & Rawlinson. The writer is indebted to Arthur J. Flint for these details.

[23]P. H. 20.11.1863.

[24]T. O. -.6.1868. According to Llandinam papers, 6, in 1866, Davies & Roberts obtained gunpowder from Curtis & Harvey of Swansea, and also Ballincolig Royal Gunpowder Mills Co. Ltd., via Charles Griffiths, Saundersfoot, agent for the Pembrokeshire district.

[25]NCB records, G. S. W. C. p. 179.

[26]NCB records, W. W. 13162.

[27]Vickerman. Letters of 2.7.1874 and 18.9.1874. The agreement between Cadman and Vickerman appears to have included provisions that the B. C. C. & I. Co. should act as vendor of Cadman's coal in the malting market, and that a proportion of coal had to be dispatched via the harbour rather than via Kilgetty station on the P. & T. R., which was much closer to Shipping colliery. In June 1874 Cadman seems to have sent his inferior coal direct to a maltster via Kilgetty station, causing embarrassment to the B. C. C. & I. Co.

[28]Vickerman. Letter of 8.11.1881. NCB plans indicate shafts at Shipping Upper, but Cadman's new pit may well have been a drift. Even in 1981 the writer was able to locate a drift at this site.

[29]NCB records. The survey may have been restricted to Shipping Upper.

[30]Letter of 25.10.1932, J. F. Vickerman to Eaton, Evans & Williams, Haverfordwest.

[31]NCB/12407 and R13542. Edwards (1950). Edwards notes that 95 men were employed at Broom in 1934, and 208 in 1938. The use of electricity and the longwall method of mining no doubt contributed to the relatively high output.

[32]P. C. 4076, 4080.

[33]Thomas Sleeman of Tenby was another proprietor. W. W. 454 and S. R. H. 19.2.1838.

[34]Edwards (1950) and (1963) p. 26, making reference to the Cash Book of the Tenby & Begelly Coal Co. The "Peggy", sloop carrying about 20 tons, conveyed coal to local ports in South Wales. The "Mary Anne", a brig of 100 tons, sailed to the south of England, often returning with cargoes of timber, bricks or limestone.

[35]S. R. H. 7.2.1840.

[36]S. R. H. 5.6.1843, P. I. C. 17.1.1853.

[37]S. R. H. 14.2.1844, 2.6.1845.

[38]G. S. W. C. p. 175/6.

[39]P. I. C. 17.1.1853.

[40]Plans and sections of the Whitland & Milford Haven Railway, 1863; G. S. W. C. p. 175.

[41]P. C. 4099.

[42]P. C. 4101, 4102.

[43]P. C. 4099.

[44]P. C. 4115 Black Hook may have been a pit of some antiquity. Morgan refers to a dispute over the mining of coal and culm at Black Hook in 1729.

[45]NCB records.

[46]T. O. 2.9.1869. Morgan assumes that Page took up the lease, for a 7 or 14 year term. This is certainly possible, but if so, a 7 year term seems the more likely.

[47]In May 1877 the Tenby Observer reported the death of J. M. P. Child, aged 52, the "only son of the late Captain Child by Emma Elizabeth Townshend", and in 1882 reported the 21st birthday of A. T. Hook-Child of Brightlingsea, "only son of J. M. Child". A search of the burial register at St. Issells has failed to reveal the date of J. M. Child's death, although a memorial tablet in the church records that his wife Emma Elizabeth died on 5.9.1861 aged 63. Morgan states that Captain J. M. Child died in[e] December, 1870 and was buried at Begelly.

[48]Mining and Mineral Statistics, H. M. S. O., 1886.

[49]The very last attempt to mine at Begelly was in the 1920's but it is doubtful if the pits concerned produced much coal. In the early 1920's a Mr. Griffiths of Begelly House Farm opened a small drift on his land, but this had a brief life. The official Lists of Mines 1926-1928, refer to Osmond Gwyther as proprietor of the Small Drink pit, Begelly. In 1927 and 1928 it was not worked at all!

[50]G. S. W. C. p. 175.

[51]P. C. 4091, 4094.

[52]P. C. 4116.

[53]See "The Story of Milford", Sir J. Rees. University of Wales Press, Cardiff (1953).

[54]NCB/4450.

[55]P. C. 4097.

[56]S. R. H. 27.2.1836 and 4.11.1836.

[57]P. C. 3826 -.

[58]G. S. W. C. p. 173.

[59]P. C. 3826-3832. Industrial archaelogists who inspected the shaft before it was capped found it to be shaped like a squashed 'D', the flat side facing the eingine house. It was beautifully constructed with dressed stone as far as the eye could see. According to W. S. Gresley's report, 9.11.1878, the shaft was oval and divided into three, the two end divisions being used for pumping and ventilation and the central one being "fitted up with double cages etc for winding coals".

[60]Lewis & Bidder's report, 1.2.1873, bears a note to this effect dated 9.1.1882.

[61]G. S. W. C. p. 173.

[62]Letter of the late Lord Merthyr to the writer, 21.4.1964.

[63]Vickerman. Memorandum 26.11.1900.

[64]NCB/4450, R13544, R13546.

[65]NCB records; Vickerman memorandum of 26.11.1900.

[66]NCB abandonment records; 4450.

[67]Industrial archaeology enthusiasts searched the tips at Lower Level colliery in 1970 and found the remains of three end tip trams, a pit prop wagon, and certain other equipment.

[68]NCB/R13546. The workings were in the Lower Level vein.

[69]O. S. Map, 6" to 1 mile, 1887.

[70]Edwards (1950) The output was very largely "slash" or small coal. It was hauled out of the pit by Ivor Howell's ponies.

[71]NCB/SW31.

[72]G. S. W. C. p. 174. The Picton Castle papers include a map by John Butcher of "The Demesne of Kilgetty belonging to Sir John Phillips, BT, 1743". This shows coal pits to the north-west of Kilgetty House and Farm in the fields known as West Hills and Malthouse Meadow.

[73]The boundary of Amroth parish and St.Issells parish passes close to the site of

Wilson's Pit, and follows the Ford's Lake stream down Pleasnt Valley to Wiseman's Bridge. Accordingly a number of the Kilgetty pits were within Amroth parish.

[74]NCB/4450, 12441, 13540, 13548.

[75]PC 1676A, 4081, 4082.

[76]PC 4092, 4095.

[77]PC 4096.

[78]PC 4086, 4114. Although small, White Leys pit is believed to have been deep. Its coal was teken down into the valley by a track which passed the property known as Pleasant Green.

[79]PC 831.

[80]Records in the Glamorgan Record Office, Cardiff, reference D/DN.A.I.-M113.

[81]SRH 3.6.1839

[82]PC 4099.

[83]W. W. 13088. Edward Wilson held a lease of coal and iron mine from Thomas and Anne Stokes in 1844/5.

[84]S. R. H. 11.1.1842.

[85]P. I. C. 29.5.1850, and accounts for the period.

[86]PC 4116.

[87]P. I. C. 27.5.1857.

[88]Lewis & Bidder's report 1.2.1873.

[89]NCB records.

[90]Vickerman. Memorandum of 26.11.1900.

[91]Edwards (1950). There is said to have been a 25′ diameter water wheel at Kilgetty colliery (Stepaside) driven by water coming from a leat off Ford's Lake and continuing into the mill race serving the Slate Mill further south.

[92]G. S. W. C. p. 172 and NCB records.

[93]L. L. 500. This survey was made for the Picton, Hook-Child, Lawrenny and Corston (Leach) estates.

[94]This account was given to the writer by W. Phillips, one of the colliers involved.

[95]Although Broom colliery and Kilgetty colliery were controlled by separate companies there was in fact a considerable overlap in direction. The directors of the Kilgetty Anthracite Colliery Co. were Lord Merthyr, J. F. Vickerman, and Trevor T. Taylor. Both pits were under the same manager—H. D. Connick in 1937, and R. J. Williams in 1938.

[96]Edwards (1950). The coal has been described to the writer as being of "very good quality and good to work".

[97]West Wales Guardian 6.1.1939.

[98]W. T. 16.2.1939; 6.4.1939.

[99]In or about the 1860's, according to a local story, a teenage girl had an illegitimate child and was disowned by her family. She was sent away to Narberth, but on her return to the district threw the unfortunate baby down Thomas Hill pit. She was arrested, but the Rector of Crunweare took up her case and is said to have defended her in court by swearing on oath that she (and all her family) were mentally defective. On the strength of this claim she was sentenced to seven years, and not given the death penalty! Thomas Hill pit remained open until the 1950's; in the late 1940's when there was still some interest in finding unworked coal, a few men descended it and managed to explore part of the adit running west towards Merrixton.

[100]NCB district map. P. C. 4099, 4101, 4102. It is believed that the Colby family of Colby Lodge near Amroth had an interest in some of the pits in this valley.

[101]PC 4084, 4085.

[102]PC 4093.

[103]W. Phillips.

[104] EEW 1573, 2175.

[105] NCB records; G. S. W. C. p. 172.

[106] Records in the Glamorgan Record Office, Cardiff, reference DNAI M135/19. The engine in question may have been used in iron ore working.

Chapter Five

THE IRON INDUSTRY

Iron Mine and Iron Mining

It is not known when ironstone (iron 'mine') was first worked commercially in the Saundersfoot area, although some such activity was probably under way before the end of the eighteenth century. It is believed that soon after Pembrey old harbour opened in 1819 it was receiving cargoes of Saundersfoot ore, and there can be little doubt that this trade was well established in the early 1820s, by which time iron mine was required for the furnaces of Gaunt's ironworks at Gwscwm. Some ironstone may also have found its way round the coast to the ironworks at Carmarthen. In the Saundersfoot district the open boats of the period were usually loaded by men and women emptying bags of iron mine directly into the vessels as they lay beached on the sand.[1]

A brief description of the ironstone, and the iron bearing veins or 'pins' has been given in Chapter 2. The ore was obtained initially from level or groups of levels known as "patches" driven into the cliffs between Saundersfoot and Amroth. Then, as now, the distinctive rusty red pins of ironstone were easy to see in the rocks, and in spite of the hazards of erosion and the tides it was natural to begin digging on the cliffs. It seems that it was not until the opening of Kilgetty ironworks in 1849 that there was any systematic attempt to develop levels inland, although at that time a number were opened, most of which were within easy reach of the ironworks.

T. R. Perkins, writing in 1934, provided an account of working methods on the patches.[2] He said that they provided employment for between fifty and a hundred men, each small section or patch being the responsibility of two men who worked together. To assist them each patch had its own little tramway, and wooden trams or trucks for the disposal of material by tipping it out onto the beach, sometimes as much as fifty feet below. The ore was then piled into heaps or dumps about 2ft. 6in. high to allow the elements to remove waste matter, although the sea often intervened as well by dragging iron mine down the beach.

According to Perkins, women and children assembled as each tide started to ebb and collected any scattered pieces of ore. These pickers usually worked as families and made their own heaps of iron mine, eventually being paid, like the miners, on a tonnage basis. In addition older women known as "pollers" were employed to sort the ore to ensure that only good pieces were used in the furnaces. Apparently they were equipped with small pointed hammers to help them check the quality and tap off unwanted shale. Duly cleaned and screened the best iron mine was taken away by boat off the beach or by farm cart to Kilgetty. According to another source, on arrival at the ironworks the mine was usually taken to calcining kilns for preliminary treatment before dispatch to the furnaces.

Perkins account leaves some scope for comment. Iron mining was to some extent a seasonal occupation. Working the patches was frequently difficult in winter, and the iron mine obtained might have to be stockpiled in the "mineyard" on the cliffs at Crickdam for weeks at a time.[3] The very problem of moving iron mine from piles at the top of the beach to the ironworks at Kilgetty probably brought about the construction of tunnels direct from Crickdam, and from the Bridge patch, through to Pleasant Valley. The dimensions of these tunnels is not known, but they are believed to have been large enough for a small tramway.[4] Certainly at the Pleasant Valley end an embankment is still visible leading from the now vanished tunnel mouth towards (but not into) the formation of the Saundersfoot Railway. Rather surprisingly there is no clear indication of a bridge over the Ford's Lake stream, although such a bridge would seem probable. If the embankment simply terminated at the stream shear legs could have been used to tranship iron mine over the water and into trams on the Saundersfoot Railway. It has also been suggested to the writer that mine may have been taken down the stream in tub boats to Wiseman's Bridge. There is no firm evidence for either of these alternatives. The tunnels were abandoned by or before the final closure of the ironworks and no obvious trace of them

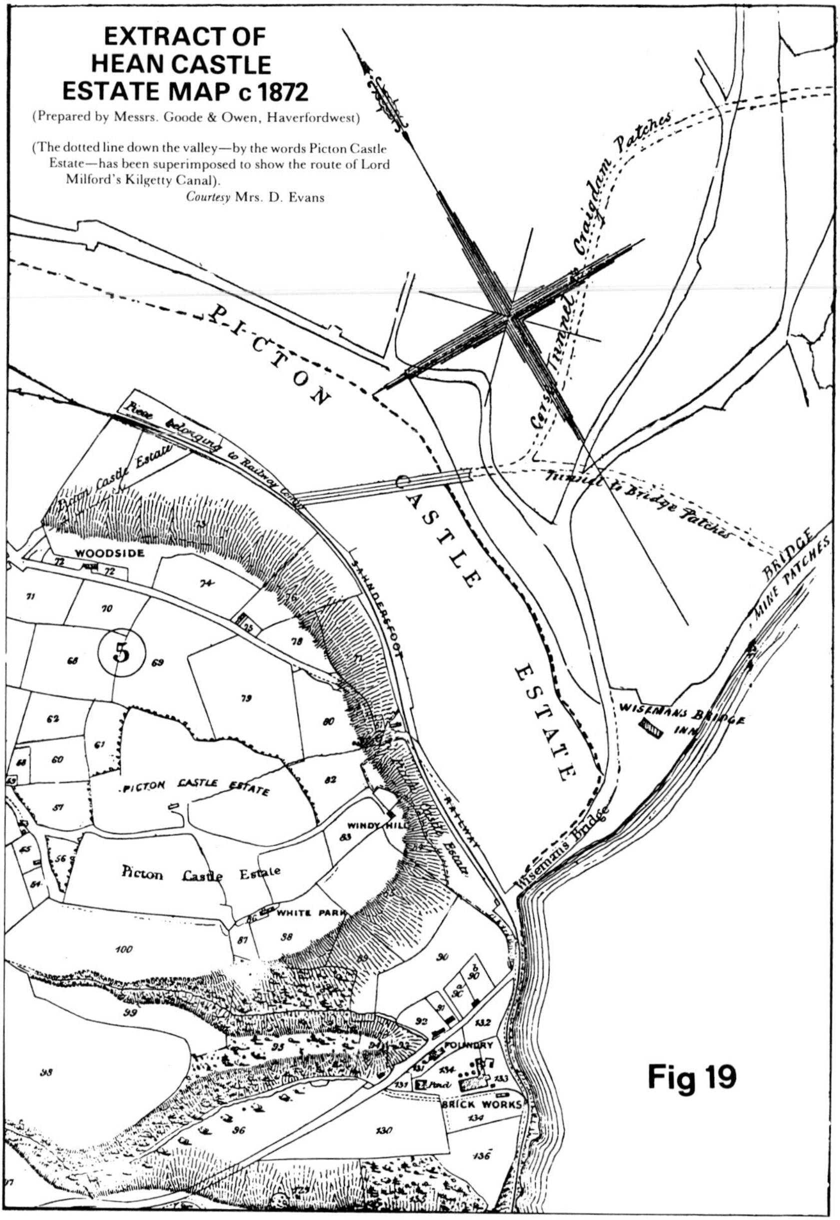

EXTRACT OF HEAN CASTLE ESTATE MAP c 1872

(Prepared by Messrs. Goode & Owen, Haverfordwest)

(The dotted line down the valley—by the words Picton Castle Estate—has been superimposed to show the route of Lord Milford's Kilgetty Canal).

Courtesy Mrs. D. Evans

Fig 19

remains amid the broken rocks of the now largely collapsed cliff patches.[5]

It is possible that before the ironworks closed the tunnels were found to be unsafe, necessitating the use of other routes to Kilgetty. The well known local writer Roscoe Howells has described how the winter stocked ore could not be removed until the summer tides permitted a cart road to be made on the beach. He has also made the point that the women and children mine pickers were required primarily when the higher spring tides spread iron ore and rubbish across the beaches. They would then gather up the iron mine into two or three ton piles by the side of the cliffs ready for movement to the ironworks.[6] If the observations of R. H. Franks on the iron industry have any relevance to Pembrokeshire, in the era of the Childrens Employment Commission, 1840-1842, girls were regularly at work on the patches. Franks noted that girls were "very frequently employed in wheeling and otherwise assisting the workmen". "The patch girls" as they were called led "a sort of half-savage life". "Hardy, and exposed to all kinds of weather", wrote Franks "they work as hard as the men from whom they differ but little in dress, and quite equal in grossness".

In the light of these varied descriptions, the organisation of labour on the patches is not precisely clear. In 1849 (and no doubt later) wages were paid by the Pembrokeshire Iron & Coal Co., the cost of labour on the patches and levels being about one third of the company's total wages bill. A Mr. Howell was evidently in charge of the P. I. & C. Co.'s work force, but it is not certain whether he was a contractor or a company servant.[7] The iron-miners and workmen, pickers and pollers appear to have been cast in the role of sub-contractors. Certainly they were paid periodically at similar rates according to the actual weight of ore they handled: their output when the ironworks was operative was often 150 tons per week, and in 1873 it fetched (after delivery to Kilgetty) 13/6d. to 14/- per ton.[8] In 1876, during a period of depression, iron mine is believed to have been worth 9/6d. per ton.[9]

The present writer has been told that the first Monday of each month was known as "Mabon's Day", supposedly a day off to allow time for reckoning up on the iron mine, making adjustments after it had been "polled", and settling payment. It has not been possible to find any documentary evidence in support of this tantalising information, and its veracity must be doubtful. There may well have been a day set apart each month for reckoning up, but it is very unlikely to have had this name. In the late nineteenth century William Abraham (nicknamed "Mabon"), a notable trade unionist and miners' M. P., won an extra day of rest for colliers throughout South Wales. His achievement was honoured in the name "Mabon's Day". This was instituted in the coal industry in 1888, several years after the last record of iron mining at Saundersfoot.[10]

Patches and Levels

Some iron mine was found in the vicinity of Swallowtree Bay at Saundersfoot, but the patches were really developed north of Coppet Hall and east of Wiseman's Bridge. On the *Hean Castle* estate between Coppet Hall and Wiseman's Bridge there were at least seven levels, six for ironstone and one for coal, and all were worked by or on behalf of the Stokes family.[11] Four of these levels were on the Catshole vein, and at least one was of very considerable extent, reputedly leading via underground workings towards an inland level at Kilvelgy. In recent years access to several of these levels has been blocked, and only one is now obvious.

East of Wiseman's Bridge the levels were even more numerous, the contorted strata of the cliffs being worked in four substantial "patches" known as Bridge, Lloyds, Crickdam and Burrows patches. The first of these, *the Bridge patch,* probably gained its name from its proximity to Wiseman's Bridge. The iron mine worked here was from a vein or pin associated with the Fiddlers' vein. It produced 25.7%-32.6% metallic ore.[12] Further east *Lloyds (or Rooksnest) patch* was named after the landowners, the Lloyd family of Laques, Llanstephan. The Lower Level coal vein appeared in this patch, and also the thinner Garland and Fiddlers veins of coal in sequence. Ironstone associated with these veins is thought to have been worked quite extensively, but the patch may have been almost as productive of coal as iron mine.

The Crickdam (or Craigdam) patch was perhaps the most important of the patches. In addition to a number of levels driven into the

Catshole vein Crickdam was provided with workshops, store houses and a "mineyard" for sorting and stacking iron mine. In the heyday of the industry there were one or two small cottages on the cliffs, but following rock falls and erosion little remains today apart from a section of crumbling wall reputed to be part of the blacksmith's shop. As at the Bridge patch the mouth of the tunnel through to Pleasant Valley has been obscured. An analysis of ore from No. 1 Level at Crickdam revealed 21.66% metallic ore.

The Burrows patch was also very productive. Two "pins" of iron mine were worked here, one lying next to the Scad vein being opened in a row of levels in the cliffs over 200 yards east of the Crickdam mineyard. Another row of levels was driven into the Catshole mine ground just beneath the brow of the cliffs. At this site the iron mine rested on the 8″ thick Catshole coal vein, and the practice was to loosen the iron mine by removing the coal first. The strata which was worked in this mine ground came down to the beach at the Crickdam mineyard. It contained metallic ore analysed at between 16.44% and 39.32%.

When the Pembrokeshire Iron & Coal Co. was formed it appears to have taken over rights to work the coastal patches in addition to driving several new inland ironstone levels. According to a P. I. & C. Co. minute of June, 1849, the levels were to be distinguished by number and name, at least for accounting purposes. It is not known if any numbering took place, but names were given.[13] The levels included the following:

Saundersfoot or Stammers Levels These were located near the shore south of Saundersfoot harbour and may well have been worked with or formed part of the coastal levels in the neighbourhood of Swallowtree Bay. One series of five small ironstone veins had a total thickness of 10″ in 7′ 4″. Another series 200 yards away produced a thickness of 10″ in 5′ 0″.

Woodside Level Situated about 600 yards north of Wiseman's Bridge and to the west of the Saundersfoot Railway, this level was driven northwards into the Kilvelgy mine measures. Kilvelgy iron mine was also found in the workings of Grove colliery nearby.[14]

Slate Mill Level The P. I. & C. Co. drove this level near the Furnace Manager's house opposite Kilgetty Ironworks. Unfortunately the Fiddlers mine measures were too irregular to be worked with profit.

Staggers Hill and Roslyn Hill Levels These were driven in 1849 to obtain "ironstone and slate mine". The Staggers Hill workings (and probably those at Roslyn Hill) were on the Catshole vein, and yielded 12.33%-31.86% metallic ore. According to a plan of about 1849, it was intended to serve these levels by a branch railway or tramway crossing the mill leat north of the Slate Mill and water corn mill. It is doubtful whether the branch was ever made.

Kilvelgy Levels and Little Kilvelgy Shafts At Kilvelgy near Lower Level colliery three or four levels were driven southwards into the Kilvelgy iron mine measures. At least one level was on the south side of the main road, and close to it. At the premises known as Little Kilvelgy, north of Lower Level, iron mine believed to have been from the Kilvelgy mine ground is thought to have been worked by two shafts situated near the north-west and south-east side of the property. The same shafts may also have been used in conjunction with coal mining.

Trewern Level and Shaft The level, 200 yards south-east of Trewern, cut the Catshole vein 30 yards in from its entrance. The level was 130 yards long, and at the end of it a shaft went down to the Catshole mine ground at a depth of 40′.

Cwms Level, Amroth Situated about 350 yards inland from the beach, this level was driven south-west on the Catshole vein, into what was an extension of the Crickdam mine ground. The workings at Cwms may actually have connected with workings from the Crickdam patch. It yielded 24.41%-30.06% metallic ore, and is believed also to have produced some coal from the Lower Level vein.

It is not suggested that this is a comprehensive list, although almost certainly it includes the most important levels. Other levels are known to have existed at Tinkers Hill, Amroth, near Cwms Level, and near Roslyn Hill in Amroth parish, where levels driven in 1849 are believed to have been known as Cornish, Machine and New Road.[15] As very little is known about some of these levels, it may well be that they were tried and soon abandoned. Good quality ironstone was also found in the workings of several collieries, including Woolsons, Grove and Bonvilles

Court. At the time Kilgetty Ironworks opened (and possibly later) Wood Level is understood to have been a source of iron ore as well as anthracite.

After transport to the ironworks, the iron mine was broken up and heated in calcining kilns situated near the top of the furnaces (see the 1872 description of ironworks plant in Appendix No. 3). Calcination had the merit of removing carbon dioxide and other impurities, and effecting a shrinkage in the material to enable a greater weight of ore to be kept in the furnaces. This in turn allowed an increased weight of pig iron to be produced each day. Unlike the blast furnaces, the kilns did not depend on any form of forced draught.

Iron mining in the middle of the nineteenth century was probably almost as spasmodic as the operating periods (or "campaigns" as they were sometimes called) at Kilgetty ironworks. Fortunately the iron miners were by no means totally reliant on the local ironworks, any more than Kilgetty ironworks was totally reliant on local iron mine. Following the discovery of the ore haematite at Penally, near Tenby, in the mid-1860's, C. R. Vickerman intended to use it at Kilgetty, but coming late in the ironworks' history it seems that little ever reached the furnaces.[16] However, in 1873 (and probably earlier) Spanish ore was imported and used at Kilgetty together with local iron mine.[17] By now, as will be seen, the chemical composition of the available ore was a matter of immense importance, and it was not unusual for furnaces to be charged with a mixture or blend of iron ores. At all events, in 1879 it was calculated that 8,300 tons of usable mine remained in the local mine measures.[18] By then, though, the ironworks had closed and mining on the patches and levels appears to have been in steady decline. In January, 1882, a note was made to the effect that "iron ore is now being exported from Saundersfoot at the selling price of 8/6d. per ton f.o.b." In October, 1883, Mr. Vickerman negotiated the sale of iron mine to the Blaenavon company, possibly drawing on existing stock. Although there may have been iron mining at a later date, no later reference to it has come to light.[19]

Limestone

The basic ingredients for smelting iron in a blast furnace are iron ore, coal (usually converted to coke) and limestone for use as a flux. The presence of all three in or near Saundersfoot undoubtedly contributed to the decision to build the Kilgetty Ironworks. The limestone was obtained principally from quarries about three miles to the north at Ludchurch and Blaencilgoed, although for many years C. R. Vickerman leased land at Gumfreston, near Tenby, apparently with a view to working limestone.[20] In 1850 quarries at Ludchurch were leased from a Mr. Probyn; later the landowner seems to have been the life tenant of the Picton Estate. Evidently there was no love lost between the quarrymen and the colliers: on pay nights fights outside the Miners Arms at Stepaside were no rarity!

The limestone was taken to the Ironworks by cart, and on arrival at least some of it was heated in one of the group of calcining or lime kilns situated near the furnace tops. Eventually most of the limestone went into the furnaces, but it is possible that the P. I. & C. Co. converted some into lime for agricultural use to give in part payment to the farmers who conveyed it to the works. However it is thought that the flows of cart traffic were more complex than this might imply. For example, Roscoe Howells has described how carters who took iron mine to the ironworks used to exchange it for culm from the nearby Grove Colliery. They would then take the culm to the lime kilns at Blaencilgoed and collect lime for the fields or limestone for the Kilgetty furnaces. In addition local timber for use in the collieries may have played a part in the traffic.

In the nineteenth century lime kilns were also to be found at Saundersfoot harbour and at Coppet Hall. Their location suggests that they were intended to treat limestone for export, the output being used almost certainly for agricultural purposes.

Kilgetty Ironworks (Stepaside)

The Kilgetty Ironworks (nowadays often referred to as Stepaside Ironworks) were set up as a result of the survey and report made by Messrs. M. & F. Foster in 1845. Asked by Messrs.

Ranken, Ford, Longbourne and Vickerman to make "proposals to open out this mineral district", they came to the conclusion that the district contained sufficient coal and ironstone to warrant the erection of four blast furnaces. They proposed that the iron-works should be situated at the foot of the railway incline, near Saundersfoot village—an idea almost certainly linked with another scheme of the moment, namely the plan to rebuild the Saundersfoot Railway to provide a feeder to the projected South Wales Railway. In order to ensure a satisfactory supply of coal to the ironworks, Messrs. Fosters recommended new workings on the Lower Level and Kilgetty veins at Kilgetty and at Thomas Chapel. It was thought that these would enable the four furnaces to produce 11,000 tons of pig iron per year for 36 years. No difficulty was envisaged as regards ironstone because, as the report said, "the ironstone of this district is very abundant and of good quality, and a large portion may be got by surface levels driven in from the sea shore".[21]

The initial reaction to Fosters' Report seems to have been entirely favourable. John Longbourne, reporting to the Saundersfoot Railway & Harbour Co. on 17 January, 1846, called the survey "very satisfactory", although he had reservations about the capital required—an estimated £36,357-10-0.[22] Sir Richard Philipps was encouraged, too, and took the preliminary step of obtaining an Act to enable him and others to grant "mining building and other leases in the County of Pembroke subject to the Uses of the Will of Richard, Baron Milford, deceased". The Act was passed in August, 1846.[23] On 15 June, 1847, Sir Richard granted a 60 year lease of mines and veins of ironstone in the parishes of St. Issells and Amroth, together with quarries and veins of limestone, to William Chadwick, W. T. Longbourne and C. R. Vickerman. He also agreed to grant new 60 year leases to work coal and culm on his estate.[24] Accordingly the Deed of Settlement creating the Pembrokeshire Iron & Coal Company was completed on 18 June, 1847, and the new company was registered on 20 July, 1847, with offices at Saundersfoot and 4 New Bridge Street, Blackfriars, London. The authorised capital was no less than £100,000 in 5,000 shares of £20 each, with powers to raise another £50,000 in loans. Some 4,360 shares had been subscribed for on the date the company was formed.[25]

The failure of the S. W. R.'s Pembroke Dock branch, and with it the failure of the scheme to reconstruct the Saundersfoot Railway, removed an important motive for building the ironworks at Saundersfoot. Although proximity to the harbour was a help, proximity to raw materials was an even more relevant consideration, and this must have influenced the decision to build the ironworks in the valley at Kilgetty. In August, 1848, construction began with some splendid ceremonial on the site. According to a local press report "the first stone of the intended works was laid by Mrs. Vickerman of London, amid the chorus of hundreds of anxious spectators and firing of cannon, which must have been heard at several miles distance". The report went on to record a dinner provided for the visitors, and subsequent speeches delivered under the chairmanship of John Longbourne. The workmen and their families were not neglected, being "plentifully regaled with bread and cheese and ale at the expense of the company. The cannon continued playing from the heights above through the entire day".[26]

In August, 1848, there were hopes that the furnaces would be lit within six months, creating employment for "upwards of 1,000 persons". The newspaper described the engineer, Thomas Hay, as "a man of superior abilities" and J. B. Hosgood, the manager, was said to be "a person of great practical knowledge, having been engaged many years in the same capacity in the Neath Abbey and other Iron Works". By January, 1849, two blast furnaces were indeed ready for operation, and it is believed that one was put into blast using coal supplied from the pits at Lower Level and Kilgetty. The company appears to have begun trading officially on 31 March,[27] and for a short while all went well. However, at the second annual meeting of the company held in London on 30 May, 1849. W. T. Longbourne, C. R. Vickerman and their friends made serious allegations of mismanagement and default against W. Chadwick and other shareholders. A row erupted, but the objectors retained the initiative. On 6 June, concerned by the state of the accounts and by a report from Mr. Hosgood, the works manager, they resolved to suspend the construction of two more blast furnaces. An accountant, William Foley, was asked to audit the books and put them in good order for the period up to 31 March, 1849. By 13 June, W. M. Neale, the company secretary, had resigned and J. D. Lowden, one of the defaulting shareholders, had made off with the company minutes and other vital documents.

By now suspicions were thoroughly aroused: there was every sign that even before the business had got into its stride the shareholders in default had been cheating those who had paid up. Very soon the defaulters became anxious to be rid of their shares with the minimum of trouble. After unsuccessful negotiations between Chadwick and W. T. Longbourne litigation ensued. Early in August the Court of Chancery ordered the return of the vital documents and Chadwick was dismissed as a trustee of the company.[27] Although large sums were still due to the company, and the ramifications of the dispute persisted for many months, the defaulters defeat meant that the company was under new management.

The board of directors now consisted of five men: W. T. Longbourne as Chairman, C. R. Vickerman, C. Ranken, J. Disney and E. Chuck.[28] A colleague J. Lawrie had died whilst the crisis was at its height. These men had the urgent task of trying to repair the company's finances. Salary cuts and other economies had been ordered in July, and the mining of iron ore discontinued. Now on 12 September, 1849, William Foley, was appointed Secretary and Accountant to the company—a sign of appreciation of the work he had already done. After a fact finding visit to Saundersfoot by Messrs. Longbourne, Chuck and Vickerman, the directors resolved that these three should act as a committee of management. In addition it was decided to make another call on shares, and to continue proceedings against the defaulting shareholders.[29]

1849 ended at a time of extreme depression in the iron trade, with prices so low that the P. I. & C. Co. decided it was not worth the effort of putting the second furnace into blast, even though it was ready in every respect. Just to add to the difficulties a gas explosion occurred in January, 1850, damaging machinery and making it necessary to blow out the one furnace in use.[30] Repairs were completed by May but in view of the depression iron making was supended and the company concentrated on working its collieries. As this course of action minimised injury to Lord Milford's interests as Landlord he was persuaded to remit claims for rents, taking only a royalty on coal and culm until iron making resumed.[31]

So far as can be ascertained, no attempt was made to recommence iron making before 1852. At that period iron ore mining was resumed, Thomas Chapel colliery was purchased, and work began in 1853 on the sinking of Grove colliery on the hillside above the ironworks.[32] These developments all provided echoes of the 1845 proposals, but sadly so far as iron making was concerned, the progress was brief. According to a company ledger, iron mine output was stepped up to reach a peak of 1,231 tons in December, 1852, but then fell away rapidly a few months later to cease in September, 1853. As the next references to iron mine are all to shipments of ore in the year 1855 it seems safe to assume the Kilgetty ironworks stopped work in 1853 after little more than a year in production.[33]

Although the evidence is circumstantial rather than certain, it seems likely that by September, 1853 the P. I. & C. Co. was once again concentrating on the collieries in the hope that they would compensate for disappointments in the iron industry. Whereas in March, 1852, the company was raising anthracite from Lower Level, Surface Level and Kilgetty (Wilson's Pit) by September, 1853, coal was being won at Lower Level, Surface Level, Thomas Chapel and both Wilson's pit and Quarry pit at Kilgetty. In July, 1855, a new shaft at New Hayes, Thomas Chapel, began to produce coal, and two months later the nearby Hacket colliery was also in production. Grove colliery became active a year later, in September, 1856, dispatching its output down an incline to the top of the ironworks' furnaces.[34]

In view of these developments it was probably no coincidence that the company's Deed of Settlement was altered by a supplemental deed on 16 May, 1853. Following the forfeiture of 1,350 shares after the 1849 row, the deed set the company's capital at £60,000 in £2 shares, with power to issue new shares to the value of £40,000. The company was also entitled to borrow £30,000 on

Engine House, Kilgetty Ironworks, 1978

"The Platform", Kilgetty Ironworks, 1978

Ruined Engine House at Grove Colliery, 1978

Ruined Blacksmith's Shop, Crickdam Patch, 1978

Lloyds (or Rooksnest) Patch, 1978

Step Cottage and old mineyard, (on right) Wiseman's Bridge, 1979

Kilgetty Colliery site, 1979

mortgage or loan, plus a sum equivalent to half of any of the additional capital if it was created.[35] At about the same time the company took the opportunity to negotiatge new leases of land and mineral rights, but if that was an indication of optimism the mood evidently changed. Although eleven men were parties to the 1853 deed, by 1857 most had withdrawn, leaving W. T. Longbourne, C. R. Vickerman and Samuel B. Oliver as the only directors.[36]

In January, 1857, these men were given fresh reason to have hopes for a happier future. In that month Mr. C. Lean, the company's works manager, reported on how he had accompanied Mr. Tregelles, formerly of the Neath Abbey Ironworks, on an inspection of the collieries and ironworks. Mr. Tregelles warmly complimented the company on the condition of its plant, and somewhat boldly expressed the belief that only gross mismanagement could prevent the company making iron at a minimum profit of £1 per ton. To this end he recommended that both furnaces be put into blast as soon as possible, with an extra engine to provide plenty of blast. He suggested that the interior of the furnaces be repaired and enlarged, and that they be given tunnel heads instead of the existing iron hoppers. He also proposed the erection of coke ovens, mine sheds and a roof to the Cast House. Much encouraged by this advice, the directors decided to resume ironmaking.[37] Extra capital was raised amongst the existing shareholders to finance these proposals, and the necessary raw materials were stockpiled in case of labour troubles. By March, 1857, local iron ore was being mined once again to meet the needs of the Kilgetty Ironworks, and towards the end of the year one furnace was put into blast.[38]

Within eighteen months the directors regretted their decision. Competition was intense, and even good quality Kilgetty pig iron was slow to sell. The annual meeting held on 31 May, 1859, heard that the iron trade was again depressed. On the brighter side, however, it was reported that No. 1 furnace was making pig iron at less cost per ton than anticipated, and that the company had contracted for enough materials to keep it in blast for almost twelve months. The directors again resolved to increase the company's capital, but S. B. Oliver was unwilling to increase his stake. W. T. Longbourne and C. R. Vickerman agreed to find the money.[39] Already they held three-quarters of the capital, and had loaned the company £27,000 at 4% on the security of the works and stock. In view of the size of their shareholding this was a somewhat curious arrangement! Not surprisingly, though, the company manager was told to practice the strictest economy. To underline the point the directors decided to abandon the costly but disappointing New Hayes colliery, which had opened only four years before. Efforts to reach the Kilgetty vein at Grove were allowed to continue, and the board decided to persist with a recent 10% cut in the men's wages, notwithstanding any strikes. On 24 June, 1859, the Pembrokeshire Herald reported that a mass meeting of the men had taken place on Kingsmoor Common, but that their protest had been peaceful.[40]

Following the death of Lord Milford in January, 1857, the estate passed into the hands of his half-brother, Rev. J. H. A. Philipps.[41] Whereas Lord Milford had been willing to help the company by remitting rents in times of depression, the new landlord was less accommodating. On 21 September, 1859, C. R. Vickerman wrote to H. P. Goode, the landlord's agent, regarding the difficulties of the P. I. & C. Co. In a powerfully argued letter he explained that the furnaces were working at a loss, and that £30,000 had been spent at Grove in the search for the Kilgetty vein. Colliery conditions obliged the company to raise coal and culm alike, but the culm represented a loss on which they paid royalties. Accordingly Charles Vickerman asked for a reduction in the royalty on coal, and its remission on culm. He proposed no changes in rents. Receiving an unhelpful reply, C. R. Vickerman wrote back emphasizing the difficulties of the company. Amongst other things, he mentioned that they were still holding almost all the iron produced since the furnace had been put into blast nearly two years earlier, and this meant a loss of interest on capital. "My friends and myself bitterly regret having made this second attempt at iron-making" he declared, adding that they were continuing only in order to use up stocks. He concluded with a thinly veiled threat: remission of the royalty on culm would make a difference, but without it the works would cease, and they would stop raising Lower Level coal, iron mine and limestone, and the landlord would suffer greatly.[42]

Rev. J. H. A. Philipps was remarkably unmoved by this plea. He said simply that he would await the finding of the Kilgetty vein at Grove. C. R. Vickerman retorted promptly that that might take twelve months, and the example of New Hayes colliery was not encouraging. He regarded the core of the matter to be ironmaking and the royalty on culm, and observed "we have indeed acted foolishly". The landlord again deferred a decision, much to the fury of Charles Vickerman. Unfortunately the subsequent correspondence is lost, but it seems that Rev. J. H. A. Philipps must have given some ground eventually because this period of ironmaking extended into 1861. In that year 4,683½ tons were shipped out of Saundersfoot, including 1,522 tons sent to Port Talbot.[43]

On 23 July, 1861, a special meeting of P. I. & C. Co. shareholders was held at the company's offices at Kilgetty. Mr. Vickerman was in the chair, and the only others present were S. B. Oliver and W. Foley, the company secretary. Mr. Foley reported that the company had been re-registered and incorporated on 5 June, 1861, under the Joint Stock Company Acts 1856 and 1857.[44] He also said that W. T. Longbourne had ceased to be a shareholder and director having transferred the whole of his shares and interest to Mr. Vickerman. As a result C. R. Vickerman held at least three quarters of the shares and apparently had assumed personal responsibility for all loans made to the company, with the exception of one advance of £4,000. Following an application from Joseph Cadman, the company agreed to lease him the Lower Level coal vein under five hundred acres of the Picton estate at Thomas Chapel.[45] Cadman soon set about the development of Shipping colliery.

The next few years in the history of the iron industry are regrettably obscure. In April, 1862, the Haverfordwest & Milford Haven Telegraph noted that the Merrixton Bottom ironworks (i.e. Kilgetty) was at a standstill, adding with wild inaccuracy that "there is a host of gaffers living like gentelmen on the profits". Not only were profits lacking, but the "host of gaffers" were too. In fact at this period S. B. Oliver seems to have withdrawn from the company leaving C. R. Vickerman as the sole proprietor. In the following year Samuel H. Blackwell was commissioned to inspect the ironworks, and when he made report in July, 1863, he advised (amongst other things) that the blast furnaces should be considerably enlarged.[46] How much rebuilding ensued is not certain, but some investment was made and at least one furnace was back in blast before the end of 1863.[47] Although some of the high quality pig iron was used locally, at the Woodside Foundry for example, most was sold for steelmaking at works elsewhere in South Wales.[48] There is also evidence for one specialised use, in saddlers ironmongery: a Walsall manufacturer used Kilgetty pig iron in place of, or in conjunction with, the more old fashioned "charcoal iron".[49] In 1864, the first full year of production, 2,933 tons of pig iron were exported through Saundersfoot harbour.[50]

After all the previous disappointments it may seem surprising that ironmaking ever resumed. The prime reason was probably that the general economic outlook was much improved, although in addition the coming of the Pembroke & Tenby Railway may have appeared to offer new opportunities for trade. At all events this campaign—the fourth—was probably the only one in which both furnaces were in blast together.[51] This may be why it has been referred to as a time of success, but arguably it was little better than before. Much of the time only one furnace seems to have been working, and in less than four years both were, to use the local expression, "on stop". On 16 January, 1868, the Tenby Observer reported that the ironworks had been in a depressed state for some time, "partly in consequence of the furnaces being burnt out and requiring to be new cased". However the newspaper added that repairs had been completed and "orders arrived to put the furnaces into blast forthwith". This may have happened, but as no other evidence for ironmaking in 1868 has been found it seems likely that if there was any activity at the ironworks it was brief. The believed closure of Wilson's Pit, Kilgetty, at about this time is an additional reason to doubt if any ironmaking actually took place.

Although the works were not functioning in 1872, by then C. R. Vickerman had plans for their sale, together with the Kilgetty

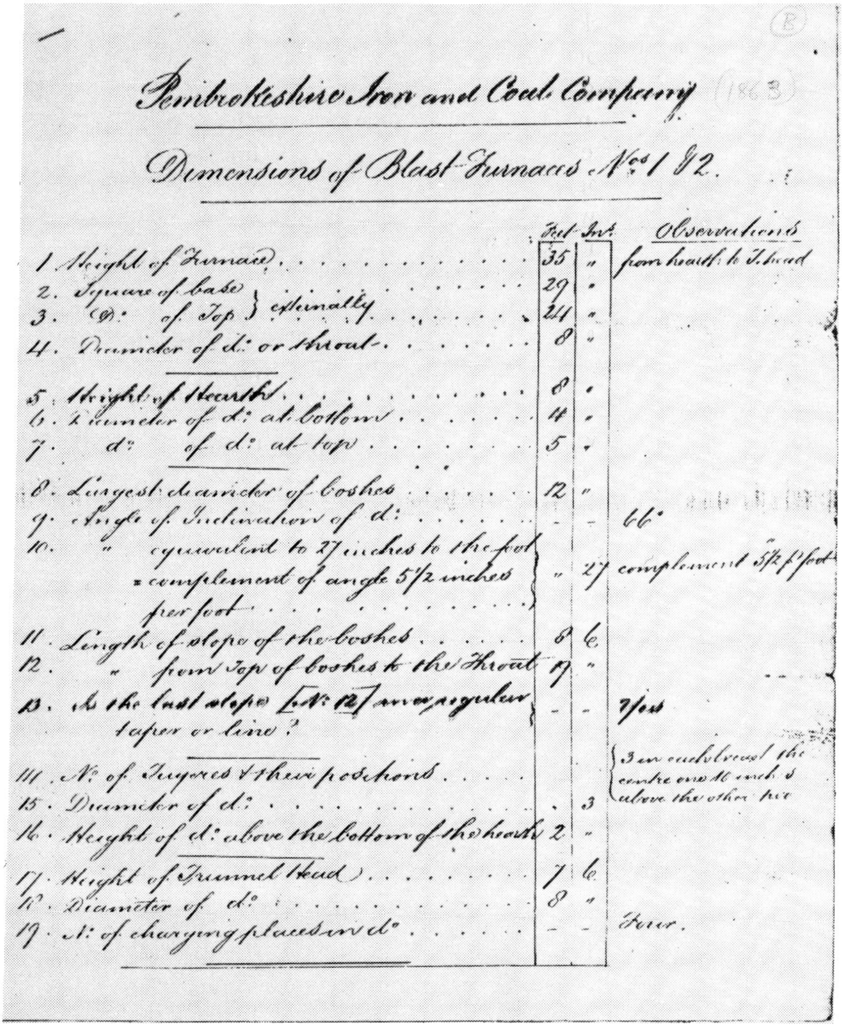

Pembrokeshire Iron and Coal Company

Dimensions of Blast Furnaces Nos. 1 & 2.

	Feet	Ins.	Observations
1 Height of Furnace	35	"	from hearth to T. head
2. Square of base } externally	29	"	
3. Do. of Top } externally	24	"	
4. Diameter of do. or throat	8	"	
5. Height of Hearth	8	"	
6. Diameter of do. at bottom	4	"	
7. do. of do. at top	5	"	
8. Largest diameter of boshes	12	"	
9. Angle of Inclination of do.	-	-	66°
10. " equivalent to 27 inches to the foot = complement of angle 5½ inches per foot	"	27	complement 5½ p. foot
11. Length of slope of the boshes	8	6	
12. " from Top of boshes to the Throat	17	"	
13. Is the last slope [No. 12] in a regular taper or line?	-	-	Yes
14. No. of Tuyeres & their positions			3 in each breast the centre one 10 inch above the other two
15. Diameter of do.		3	
16. Height of do. above the bottom of the hearth	2	"	
17. Height of Tunnel Head	7	6	
18. Diameter of do.	8	"	
19. No. of charging places in do.	-	-	Four.

Extract from S. H. Blackwell's Report, 1863 *Courtesy Mrs. D. Evans*

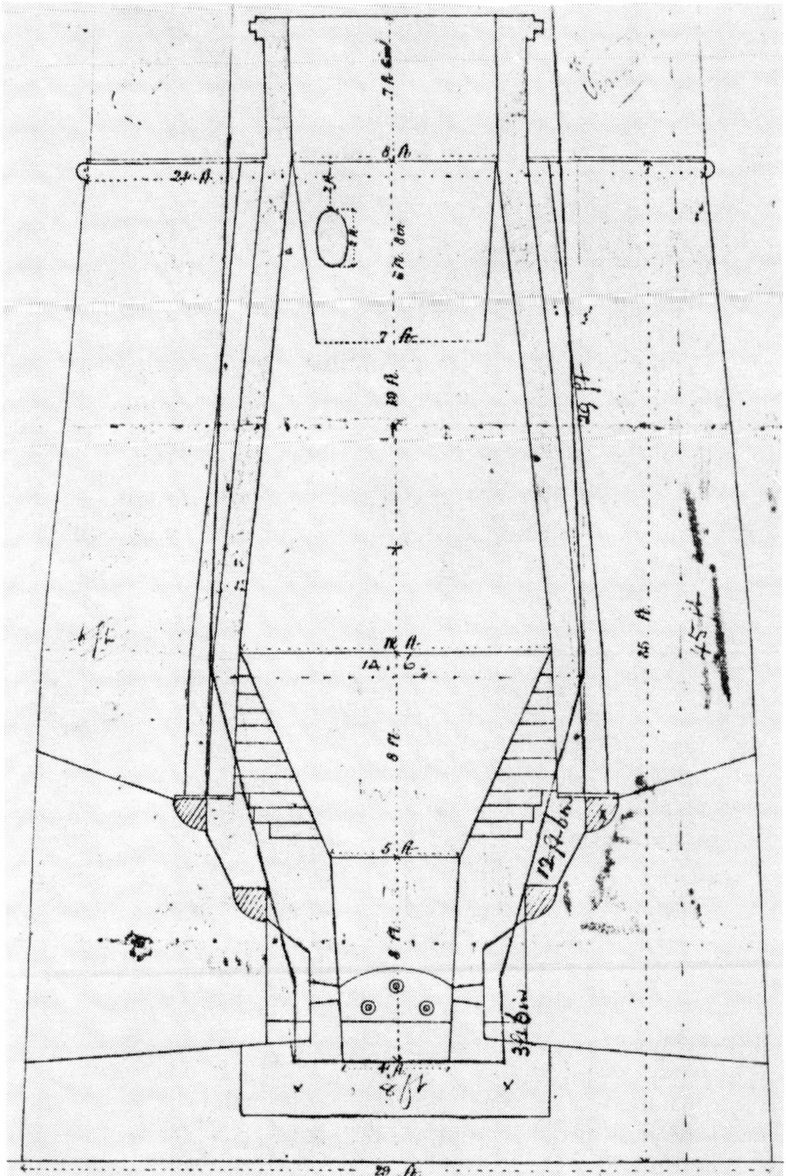

Cross Section of Furnace

S. H. Blackwell's Report, 1863

collieries, Bonvilles Court colliery and the Saundersfoot Railway & Harbour Co. Prior to the formation of the Bonvilles Court Coal & Iron Co. detailed reports were obtained from Messrs. Bird & Co. of London and Messrs. Lewis & Bidder of Aberdare. A copy of the latter has survived, and although it gives an agent's or valuer's view of the works, it mentions two interesting economies employed at Kilgetty. The first, which was as relevant to the collieries as to the ironworks, was the use of culm to heat steam engine boilers, thereby saving large coal for sale and shipment. This was achieved by the expedient of sprinkling "clay-wash" over the culm, a process which at one time was thought impracticable but in practice proved to be simple and inexpensive. It is perhaps worth noting in passing that the traditional method of making a domestic fire involved mixing culm with clay and water to make into "balls" for burning. These could provide a good and reliable heat for the local "clom" cottage all through the year.

The second economy was the manufacture of anthracite coke from culm mixed with the slack of bituminous coal and a proportion of pitch. This was said by Lewis & Bidder in 1873 to be "a first-class Furnace Fuel, seeming to unite the good qualities of both coke and anthracite coal in furnace operations". They claimed that Kilgetty was the first ironworks to employ the idea successfully, but admitted there were other contestants for the honour. Undoubtedly there had been technical difficulties in using anthracite as a fuel for smelting, and in 1838 George Crane of Ynyscedwyn Ironworks, Ystradgynlais, had been the first to claim a solution. However both his methods and the fuel he produced may not have been fully comparable with those later used at Kilgetty.[53] So far as Kilgetty was concerned the economy of anthracite coke was not only original, but also of the utmost importance to the entire works in providing a simple and effective way of disposing of the less saleable culm. Coppée coke ovens were installed to assist the process, and these are reputed to have been the first of the type in the country.[54]

The Bonvilles Court Coal & Iron Co. Ltd. was set up in May 1873, with the intention (amongst other things) of restarting the ironworks at Kilgetty. As the prospectus (reproduced in Appendix

No. 1) shows, the company's possessions included the property of the old Pembrokeshire Iron & Coal Co. No time was lost, and on 29 May, 1873, the Tenby Observer reported that "great preparations are being made at Kilgetty ironworks which are intended to be put into blast early next week. Several cargoes of Spanish iron ore have been discharged, and also a great quantity of bitumenous culm for coke." Several furnace firemen had also arrived, and there seemed to be every prospect of a revival of trade. Ironmaking duly began, although probably only one furnace was at work.

This modest scale of operation was unexpectedly short-lived. Confronted with the unpleasant reality of a deepening economic depression, the management imposed a 10% cut in the colliers' wages to take effect on 30 May, 1874.[55] Protests by the men caused the cut to be postponed for a fortnight, but in spite of some progress in negotiations the men went out on strike on 16 June.[56] On 29 June, they returned to work, many having accepted a 5% reduction—but by then the furnace at the ironworks had been extinguished.[57] So far as the present writer can ascertain it was never re-lit. Beset by labour troubles and the difficulties of depression, the collieries of the B. C. C. & I. Co. continued to raise anthracite until the second half of 1876. That autumn another dispute over wage reductions brought every pit to a standstill, and by December the company had decided to close down a large part of their undertaking permanently.[58] In 1877, following action by the company's creditors, the Court of Chancery put the company in the care of a Receiver, Henry Batchelor.[59] In the circumstances there was little hope for the ironworks' future.

Kilgetty's problems were not merely those of economic depression and industrial unrest. By now new technology was running ahead of the ironworks' resources. The production of large quantities of mild steel using either the Bessemer or the Siemens processes had become a practical proposition, and it was hardly possible for any ironworks to remain in business simply by smelting iron ore and selling it as pig iron.[60] Accordingly the later part of the nineteenth century saw a spate of South Wales ironworks closures, and Kilgetty was only one amongst many works unable to move with the times. One factor in this trend was the unsuitability of most Welsh iron ore for use in Bessemer steelmaking: the ore had a relatively high phosphorous content making it unsatisfactory for use in the Bessemer process, which initially was employed to make an acid steel.[61] Although Saundersfoot iron mine was used in the process at more than one works, the ore was probably almost as phospheric as elsewhere.[62]

C. R. Vickerman was well aware of the difficulties confronting Kilgetty and in 1878 and 1879, after his reacquisition of the assets of the B. C. C. & I. Co. (described in Chapter One), he sought expert opinion on the state of the works and the remaining iron mine.[63] At the time he had a resale of the property in mind, but the fact that problems were general doubtless contributed to the difficulty he encountered in finding any purchaser. His advisers certainly showed more interest in the collieries than the ironworks, but Mr. Vickerman was not a man to give up easily. In January 1882, he consulted David Evans of the Rhymney Ironworks, saying that Kilgetty pig iron had been well thought of at Rhymney since 1863, but that most of it had been sold to the ironworks at Dowlais for use in the Bessemer process.[64] After further advice, Vickerman favoured the idea of installing a Bessemer converter at Kilgetty, apparently on the grounds that this would enable the output of both furnaces to be run into steel ingots without the cost of reheating.[65] In the event the idea was never acted upon: after eight years of idleness Kilgetty Ironworks had lost its markets and all prospect of recovery. Perhaps reluctant to put an end to the venture, it was not until 1887 or 1888 that C. R. Vickerman decided to dismantle the works. The task was carried out by or before January 1889.[66]

Figure 16 (on page 114) shows the layout of the ironworks in 1887 shortly before dismantling began. It seems that initially only the machinery was scrapped. The large stack on the hillside above the furnaces was not felled until about 1909, when some of the usable bricks were cleaned and resold locally. The engine house and furnaces were allowed to decay, but the building known as "The Platform" remained in use for repair shops and offices until the late 1930's. Much of the corrugated iron roof of the Cast House remained in position until requisitioned for civil defence purposes in World War II. Today the surviving buildings are so crumbled and overgrown that residents in the surrounding caravans could be forgiven for not realising that they are sitting on the site of an ironworks. At the time of writing, however, these remnants of Kilgetty ironworks linger on.

The Woodside Foundry, Wiseman's Bridge

The foundry was set up in about 1850, and probably owed its development to the opening of Kilgetty Ironworks. Established on the Hean Castle estate adjacent to Thomas Stokes' brickworks, the site was held under lease in 1851 by David Parsell of Wind Hill, Saundersfoot. He in turn had a tenant, William Thomas Harries.[68] Although Parsell was an engineer his role seems to have been that of an inventor and perhaps financial backer; Harries occupied the foundry and apparently managed it. Pieces of ironwork survive bearing the stamp "W. T. Harries, Woodside Foundry, near Tenby".[69]

By 1864 Parsell had a partner in W. J. Hesketh,[70] and by 1871 he had a business association with S. S. Williamson of Moreton colliery.[71] With Williamson's co-operation Parsell developed a number of engineering ideas, primarily for use in the coal industry. At least three of these were patented, earning Parsell the nickname of "the George Stephenson of the Saundersfoot district".[72]

In November 1866 Harries supplied cast iron railway chairs to the newly opened Pembroke & Tenby Railway.[73] At the time the Foundry advertised that they would make "all kinds of castings for collieries, railways, gas and water works on the shortest notice. Lowest price for castings. Old Brass and Iron taken in exchange". It is not known when Harries vacated the Foundry, but he handed it over to a nephew with the surname of David. Subsequently the Foundry was run by Thomas David in partnership with John and Joseph David. Thomas David, the last of the line, died in 1922 and his widow tried to keep the foundry at work in order to sell it as a going concern. Unfortunately this arrangement did not succeed, and the Foundry had to close in 1924.[74]

In the early years the foundry used pig iron from Kilgetty, but after 1874 it had to be imported from elsewhere in South Wales. A cupola furnace was used for the manufacture of a wide variety of industrial castings ranging from colliery tram wheels to propellors for coal boats. The foundry also turned out domestic ovens and boilers, gates and occasionally even cast iron tombstones, examples of which can still be seen in the graveyards at Sardis and St. Issells.[75] There is no record of work in wrought iron being carried out, but casting was sometimes done in brass.

The highly skilled task of pattern making occupied a lot of time at the foundry, but a casting day was regarded as a local event. Children would hurry home from school, and gather at the foundry to watch the sparks fly from the molten metal. If they indulged in pranks like lobbing pebbles into molten metal they were rapidly rebuked. Even so, over a half a century after the foundry's closure the David family is still remembered with real affection.

Footnotes

[1]Howells p. 34.

[2]T. R. Perkins, "The Locomotive", 15.9.1934.

[3]PC 4116 shows that the output of iron mine actually reached a peak in December 1852, (see later in this chapter) although there is no indication as to how much mine was won from the patches and how much from the inland levels.

[4]NCB 13540. This refers to "the Corse tunnel for taking iron mine from the Crickdam patches and levels to the Saundersfoot Railway." The other tunnel is described as "branch tunnel to the Bridge patch". The stone arched tunnel mouth at the valley end has been sealed within living memory.

[5]NCB 13540 and two other maps marking the tunnels either appear to show that the stream was not bridged, or do not show the arrangements at all. The remains of the embankment now stops short of the stream, but there is just enough stone to be seen on each side of it to allow for the possibility that originally the stonework was part of bridge abutments. It is believed that a mineyard once existed just to the north of Step Cottage, Wiseman's Bridge, alongside a siding on the Saundersfoot Railway. This would fit in with the tub boat theory, and yet the embankment in the valley was clearly built to meet the route of the railway further north, closer to the ironworks.

[6]Howells p. 35. The present writer has been told that the landlord of the Wiseman's Bridge Inn had a fine pair of horses which he used to haul iron mine from the patches to the Saundersfoot Railway at Wiseman's Bridge. If this is correct it would appear to confirm the existence of a mineyard at Step Cottage and suggests that the tunnels from the patches did not have a long or successful career.

[7]PIC.

[8]Lewis & Bidder's report 1.2.1873.

[9]Howells, p. 35.

[10]E. D. Lewis: "The Rhondda Valleys" Phoenix House, London (1958) p. 166 et seq.

[11]G. S. W. C. p. 164.

[12]PC 4099.

[13]PIC 6.6.1849.

[14]Vickerman. The memorandum of 26.11.1900 refers to expenditure on driving the Woodside Level in the period 1878-1889, but no other mention of this work has been found. At this time the intention was almost certainly to win coal, but the very lack of information suggests that it was not very successful.

[15]NCB records and 13540.

[16]Lewis & Bidder's report 1.2.1873.

[17]T. O. 29.5.1873.

[18]Vickerman. Memorandum D. John to C. R. Vickerman 4.4.1879.

[19]Vickerman. Note of 1.10.1883.

[20]PC 4101, 4102. In 1889, following further financial loss, C. R. Vickerman surrendered his lease of 178 acres at Gumfreston.

[21]PC 4099.

[22]S. R. H. 1.6.1846.

[23]PC 2029, 2030, 4562, 4563.

[24]Vickerman: P. I. & C. Co. Deed of settlement.

[25]Vickerman: P. I. & C. Co. Deed of settlement.

[26]PH 11.8.1848.

[27]PIC 6.6.1849.

[28]PIC 8.8.1849.

[29]This board took over at a meeting on 27.6.1849. The inclusion of E. Chuck of Ware, Herts, is noteworthy: he was a prominent maltster. Bridges Harvey, C. R. Vickerman's brother-in-law, replaced J. Disney on 1.9.1852. PH 17.8.1849 reported that the company had "discharged about 200 persons from the Ironworks and 30 from the Coal Works owing, it is supposed, to the depressed state of the iron trade". The furnace in blast was said to make "upwards of 60 tons per week".

[30]PIC 8.11.1849.

[31]PIC 29.5.1850.

[32]PC 4116. P. I. C. records make no direct reference either to the resumption of ironmining and ironmaking, or to the sinking of Grove colliery. However a balance sheet for 1852/53 includes items of expenditure on "Labour in obtaining mine" and "Labour at furnaces".

[33]PC 4116.

[34]PC 4116.

[35]Vickerman: Supplemental Deed, 16.5.1853.

[36]P. I. C. 21.1.1857.

[37]P. I. C. 21.1.1857.

[38]P. I. C. 27.5.1857 and 16.9.1857.

[39]P. I. C. 31.5.1859.

[40]PH 24.6.1859.

[41]Rev. J. H. A. Gwyther, Vicar of Madeley, Shropshire, and half-brother of Lord Milford, was licensed to take the name and Arms of Philipps (PC 2091, 2093, 4797).

[42]PC 3826.

[43]Edwards (1963) p. 22.

[44]PIC 23.7.1861.

[45]PIC 23.7.1861.

[46]A plan of the blast furnace improvements appears to be the only part of Blackwell's report to survive.

[47]Edwards (1963) p. 22.

[48]Vickerman: Lewis & Bidder's report 1.2.1873.

[49]Vickerman: 1882 note endorsed on Lewis & Bidder's report.

[50]Edwards (1963) p. 22.

[51]Edwards (1963) p. 22. According to Lewis & Bidder the furnaces together could makc up to 300 tons of pig iron in a week, but the very few figures available do not suggest that this was achieved often, if at all. However the 1872 inventory of ironmaking plant (see Appendix No. 3) makes it clear that both furnaces (and No. 1 in particular) were used.

[52]Lewis & Bidder's report, 1.2.1873.

[53]Morris & Williams, p. 44.

[54]Coppée ovens are thought to have been built of iron. Those at Kilgetty were erected under the guidance of M. Soldenhoff, of Belgium.

[55]T. O. 11.6.1874.

[56]T. O. 11.6.1874.

[57]T. O. 25.6.1874.

[58]T. O. 7.12.1876. At this date the men had been on strike for two or three months following a proposal to cut their wages by 15%.

[59]Vickerman v Bonvilles Court Coal & Iron Co. Ltd. and others. P. R. O., Chancery Lane, Ref. J15/1337. H. Batchelor was the secretary of the company prior to liquidation, at which time W. Foley was works manager. Foley, who had been associated with the business since 1849, left the district in 1878.

[60]The Bessemer process involved rapidly purifying liquid pig iron by blowing air through it in a vessel called a converter. The Siemens process was associated with the open hearth type of furnace, not used at Kilgetty.

[61]Most U. K. iron ores contain a good deal of phosphorous and are therefore better for making basic steel. The cost of importing ores low in phosphorous and other impurities was a real problem for many S. Wales ironworks.

[62]Unfortunately no chemical analysis of Saundersfoot iron ore is available.

[63]Vickerman: W. S. Gresley's report 4.11.1878; Report on Hean Castle minerals, 24.3.1879; D. John's memorandum on iron mine 4.4.1879.

[64]Vickerman: Letter to D. Evans 31.1.1882. Dowlais Ironworks was actually the first in Britain to adopt the Bessemer process.

[65]Vickerman: Memorandum of 9.1.1882.

[66]Vickerman: Note endorsed on memo of 9.1.1882.

[67]Reported recollection of the late Lord Merthyr.

[68]WW 13094. The terms of the lease of 1.12.1851 indicate that the foundry was associated with Woolsons pit. A remelting cupola with blowing machinery "now attached and situate at Woolsons work" was included in the lease, Parsell to pay rent for it at 2½ % on sales of castings from the cupola. He had free use of the weighing machine, blacksmiths tools, bellows, anvil etc.

[69]Some were included in the exhibition on Saundersfoot's Industrial Past opened at the Kilgetty Information Office in 1976.

[70]Plans of the South Wales and Milford Haven Junction Railway, 1864.

[71]See the section on Moreton colliery in Chapter Three.

[72]T. O. 30.4.1874.

[73]Llandinam papers, 6.

[74]Information provided by Mrs. M. Acors, granddaughter of Thomas David. Osborne Evans, a blacksmith at the foundry at the time of closure, has advised the writer that the business was sold to Messrs. Hancocks of Milford Haven.

[75]Stickings p. 50. The foundry employed about a dozen men, one of whom, the blacksmith, Dick Absalom, served as clerk to Amroth Parish Council for 44 years.

Chapter Six

TRANSPORT

A Early Transport and Transport Schemes

In the eighteenth century the people of Pembrokeshire relied on two forms of transport for the movement of goods. One was the small sailing boat, and the other was the cart, drawn either by horses or oxen. The importance of coastal shipping can hardly be underestimated in an area where no one lived very far from the sea.[1] At this time it was probably easier to travel to Swansea or Bristol, or harbours in North Devon and Cornwall, than to travel inland to, say, Brecon or Hereford. To judge from a surviving record of coal shipments from Wiseman's Bridge in 1789, boats regularly sailed the length of the Bristol Channel to destinations as unlikely in the present day as Frampton and Berkeley, Gloucestershire.[2] Even after the railways reached South Pembrokeshire the sea remained an important means of transport: in 1867 the local press advertised passenger sailings between Bristol and Tenby by the Bristol & General Steam Navigation Co.'s "powerful steamers", the "Briton" and the "Undine".[3]

The state of the local roads was appalling. Few (if any) were properly surfaced, and the traveller had to contend with deep mud in winter and swirling dust in summer. In the eighteenth century horses seem to have been in short supply, and oxen were often used in West Wales, and not only on the very inadequate roads. In about 1766 Thomas Kymer of Robeston Hall, Robeston Wathen, near Narberth, used oxen on his wooden railways in the Carway area of the Gwendraeth valley. He also employed a man from Pembrokeshire who was skilled in the craft of shoeing oxen (a "clewer"). On this basis it seems reasonable to suppose that similar operations may have existed in Pembrokeshire.[4]

The later part of the eighteenth century saw the rise of the turnpike trusts, maintaining main roads to some sort of standard in return for tolls paid—but it was many years before this development had any relevance to the Saundersfoot district. Ironically the most notable improvements made were motivated by the needs of the Royal Navy. In 1814 the government transferred the naval dockyard from Milford to Pembroke Dock, then often known as Pater. Communications with both sides of Milford Haven remained woefully poor, and eventually a select committee of the House of Commons looked into the matter. As a result Thomas Telford was asked in 1827 to carry out a survey of the main overland route to the Haven. Not only did Telford advocate the amalgamation of existing turnpike trusts but he also recommended the construction of a new and more direct piece of road to Pembroke Dock, where he favoured Hobbs Point as a place of departure for Ireland. The main portion of this new route ran from a point near St. Clears, through the village blessed with the name of Red Roses, and on to Killanow, Kilgetty and Kingsmoor. At Kingsmoor it met an existing road maintained by the Tavernspite Trust, and this trust was responsible for building another part of the route running west from Kingsmoor towards Broadmoor, Redberth and Pembroke Dock.[5]

It is sometimes said that Thomas Telford actually built the new road, but it must be doubtful if he did much more than conduct the initial survey, and thereafter act as consultant. At this period he was at the pinnacle of his profession, and so much in demand it seems unlikely he could have spent much time in West Wales. Furthermore Telford was not young, and he died in 1834 whilst the new road was under construction. Building the new road occupied much of the 1830's, and in accordance with Telford's methods it was no doubt given substantial foundations, carefully laid and surfaced. The principal engineering work on the route was the bridge across the valley at Stepaside. This fine structure survived intact well into the age of the juggernaut lorry. In 1978/1979 it was partially rebuilt, and the twin arches in the bridge—one for a minor road and one for the Saundersfoot Railway—were replaced by a single concrete span.[6] In its original condition the bridge (and the new road) were first used by the mail coach on 6 April, 1839.[7]

No reference to turnpike trusts in Pembrokeshire can be made without some mention of the Rebecca riots of 1839-1844. The setting up of extra toll gates in the late 1830's created serious problems for small farmers already struggling to cope with economic depression. They expressed their outrage by attaking and destroying the gates, often evicting toll keepers and their families in the process. The mob was led by Rebecca or "Becca", a man dressed in female clothing. Some of the most serious incidents happened in the vicinity of Whitland, but the new road from Carmarthen to Pembroke Dock was not exempt. Order was not restored until the local constabulary received outside assistance, and a commission of enquiry was set up to look into the grievances.[8]

It is now necessary to turn to the earliest schemes specifically intended to improve transport for industrial Saundersfoot.

The Kilgetty Canal

The Kilgetty Canal was built in about 1792 to link Lord Milford's collieries at Kilgetty with the coast at Wiseman's Bridge. According to Hassall, who was writing in 1794, the canal was actually planned by Lord Milford and constructed by his own colliers. This combination of talents was not very inspired because the canal was built from the wrong end. Hassall wrote:

"It seems to be universally agreed among the skilful in the art, that canals near the coast, intended to reach a certain point inland, should begin at the highest level the ground will admit of, in order to reach as far as possible upon one line. Unfortunately the construction of Lord Milford's canal does not accord with this established principle; and is got so much under level before it arrives at the colliery it was meant to serve, that the benefits it would otherwise have produced, are in a great measure lost."[9]

So far as can be ascertained, the canal ran for little more than a mile through Pleasant Valley, terminating near the Carmarthen-Pembroke road at Stepaside. Almost certainly coal had to be taken by cart to the head of the canal, which thereby largely failed to provide the intended colliery to coast service. However it is not known to what extent the canal was actually used, if at all. But some attempt to operate it probably was made, if only to discover its unsuitability. An inspection of the canal's remains by an experienced industrial archaeologist in the early 1960's suggested that the canal had been 10′-12′ wide, without locks, and capable of accommodating tub boats only. The water supply appeared to

Ruined Engine House, Lower Level, 1979

Old Engine House at Hacket, 1979

Site of Broom Colliery, 1978

Saundersfoot Harbour entrance, 1978 (Hean Castle on hillside beyond)

Bonvilles Court Pithead, 1978

Former Bonvilles Court Coal Co. cottages, Stepaside, 1979

Spoil Tips at Moreton Colliery, 1979

Cleared Site of Broadmoor (Greenhill) Colliery, 1979

come from a stream disgorged from several adits or levels near the site of Band of Hope colliery at Kilgetty.[10] A water course passing through a well made stone arched channel is still to be seen in this vicinity, and this may have served the canal or, perhaps, a water-wheel at Kilgetty colliery.

Today the surviving traces of Lord Milford's canal have to be searched for to be found. The Saundersfoot Railway was built over the canal's formation for some distance south of Stepaside, but the two diverged where the Saundersfoot line crosses the valley from the east side to the west. About 250 yards of the old canal can be seen in the neighbourhood of Star Gorse Farm, but as the route runs south it becomes no more than a damp and barely discernible strip in the fields. At Wiseman's Bridge the course of the canal is marked only by a small ditch, most of the site being filled in and used as a caravan park. The small canal basin at the top of the beach has long been covered with shingle.[11]

The Kilgetty canal was the first and last attempt at canal building in the Saundersfoot district. In 1815 Walter Davies advocated the construction of a canal between Cresswell Quay on Milford Haven and Saundersfoot, a distance of approximately six miles. He argued that this "could not fail of highly improving the value of the collieries in that tract of the country", but nothing came of the proposal.[12]

Early Tramway Proposals

The Kilgetty Canal was closely followed by the first proposal for a tramway. As mentioned in Chapter 3 the tramroad expert John Curr wrote in 1797 to the proprietors of Moreton colliery to advocate the construction of a tramway upon the principles he had devised.[13]

Nothing was done, and after a visit to the district in 1806 Dr. Charles Collins of Swansea noted that there were no railroads or canals in use. In the same year John Curr repeated his proposal for a tramway, linking it to the suggestion that several coalowners in the area should combine to form a major concern to work the deeper seams.[14] So far as is known nothing happened, although without the backing of Richard, Lord Milford as the major landowner it is likely that little could be achieved. Certainly there is no further evidence of a tramway project until Lord Milford's death in 1823. According to a document in the National Library of Wales the trustees and tenant for life Lord Milford's estates then granted the Begelly coal owners Sarah Child and James Mark Child a licence to lay down a tramroad from their collieries over estate land to the sea shore at Saundersfoot, and power to build storehouses there. The curious feature of the document is that although it is dated 31 December, 1824 and sealed it is not signed. Not surprisingly, perhaps, no evidence has been found that the powers grantd by the licence were ever exercised.[15]

One tramway was built to link pits at Stonecross and Cants, near St. Issell's church, with the beach at Coppet Hall. No records of its construction and operation have come to light, but the line predated the Saundersfoot Railway. In all probability it did not have a long life because it is depicted as an "old tramway" on a plan dated c.1845 which now hangs in the Hean Castle estate office. The same plan indicates that the line was laid entirely on land then belonging to the Hean Castle estate—surely an indication that the tramway was built, owned and operated by the Stokes family, the occupiers of Hean Castle in the first half of the nineteenth century, or by John Wogan, owner of Hean Castle in the late eighteenth century. It has been said that the line was properly laid with sleepers, and worked by horses, but these assertions must be regarded as likely rather than certain.[16] Today much of the route of the tramway is used as a footpath called the Black Walk, so that the line is now sometimes referred to as the Black Walk Tramway.

B The Saundersfoot Railway

Formation and Construction

The Saundersfoot Railway and Harbour Company was proposed in 1828 for the purpose of making a railway link betweeen collieries on the late Richard, Lord Milford's estate, and a new harbour to be constructed at Saundersfoot. William Bevan & Son,

Engineers of Morriston, Swansea, were employed to make a survey, and they completed their task in November, 1828.[17] Their estimate of the cost was as follows:

	£.	s.	
For the main line (to Thomas Chapel) 4¾ miles	6491	10	0
Ridgeway branch	818	11	10
Wiseman's Bridge branch	2131	9	8
	9441	11	6
Beside above prices, the price of land	525	0	0
Railway road over piers	240	0	0
Erecting the breakwater at Saundersfoot	7007	18	10
Making altogether	17214	10	4

A Bill for the construction of the railway and harbour was brought before Parliament in the next session: the preamble named Sir Richard Philipps, Charles Ranken and William Evans as the principal subscribers. The committee of the House of Lords considering the Bill met on 22 May, 1829, under the chairmanship of the Earl of Shaftesbury. They noted that the estimate of expense was now £17,364-12-3½, and that over half of the £17,500 subscribed had been promised by Sir Richard and Charles Ranken. A contract made on 8 May between them and the engineer William Aldridge Bevan was produced for the benefit of the committee, and they approved the Bill without amendment. The Saundersfoot Railway & Harbour Company's Act of incorporation duly received the Royal Assent on 1 June, 1829.[18]

Under the Act the company was given five years to complete its works, and these included the construction of the two branch lines —one from Ridgeway near Saundersfoot to Merton (or Moreton) and the other from Saundersfoot along the coast to Wiseman's

Bridge. The Act specifically stated that the company should not start building the Wiseman's bridge branch "until such time as the Harbour at Saundersfoot shall have been made and rendered fit for the reception of ships and vessels of the burthen of twenty tons".[19] This provision was probably intended to minimize interruptions in coal shipments, ensuring that once construction began those who shipped from the beach at Coppet Hall and Wiseman's Bridge would have some satisfactory alternative facility. In spite of its title the Saundersfoot Railway was conceived as a horse-worked tramway rather than a railway, and resembled a turnpike trust to the extent it allowed numerous people to operate their own rolling stock in return for the tolls and charges they paid.

Although most of the capital was already subscribed, a prospectus was issued on behalf of the company on 5 May, 1830, and signed by Messrs. William Bevan & Son, and Messrs. Goodeve and Ranken, Solicitors, of 4 South Square, Grays Inn, London. Not surprisingly, the prospectus was optimistic. It envisaged the movement of 10,000 tons of coal per annum from the Thomas Chapel collieries, 12,000 tons of coal and 18,000 tons of culm from the Begelly and Kingsmoor collieries, 20,000 tons of coal and culm over the Ridgeway branch, and another 20,000 tons over the Wiseman's Bridge line. Assessed on the basis of 2½d/ton/mile (or 2d in the case of Thomas Chapel collieries) plus 1d/ton shipment costs at the harbour, it was reckoned that the total annual revenue might total £2,900.[20]

This prospectus provides the last reference to William Bevan & Son; what happened as regards their involvement is a mystery, but by the time of the first minuted meeting of the company, held at Haverfordwest on 16 May, 1831, Mr. R. W. Jones was acting engineer. On this occasion he advised Sir Richard Philipps and William Evans that he had begun the formation of the railway near Kingsmoor, and near the intended harbour at Saundersfoot. He was asked to proceed with the works "with all possible expedition", and authorised to deal with local landowners for the land and premises required. Accordingly construction went ahead, and in January, 1832, it was agreed that Mr. Jones could contract for the necessary rails. At the same time William Longbourne was appointed Clerk to the company; at a meeting in the previous

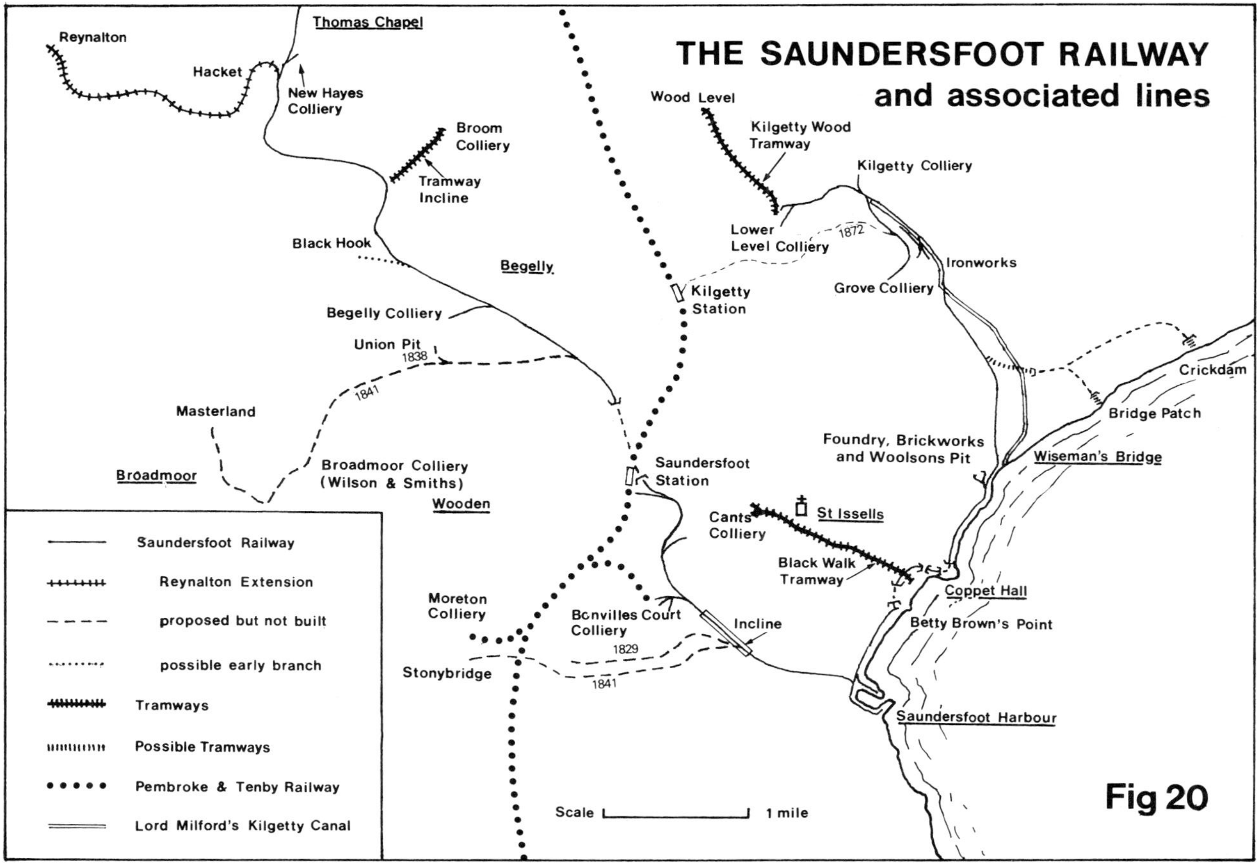
THE SAUNDERSFOOT RAILWAY
and associated lines
Reynalton
Hacket
Thomas Chapel
New Hayes Colliery
Broom Colliery
Tramway Incline
Black Hook
Begelly
Begelly Colliery
Union Pit
1838
1841
Masterland
Broadmoor
Broadmoor Colliery (Wilson & Smiths)
Wooden
Moreton Colliery
Stonybridge
Bcnvilles Court Colliery
1829
1841
Incline
Saundersfoot Station
Kilgetty Station
Wood Level
Kilgetty Wood Tramway
Kilgetty Colliery
Lower Level Colliery
1872
Ironworks
Grove Colliery
Crickdam
Bridge Patch
Wiseman's Bridge
Foundry, Brickworks and Woolsons Pit
St Issells
Cants Colliery
Black Walk Tramway
Coppet Hall
Betty Brown's Point
Saundersfoot Harbour
Scale
1 mile
Saundersfoot Railway
Reynalton Extension
proposed but not built
possible early branch
Tramways
Possible Tramways
Pembroke & Tenby Railway
Lord Milford's Kilgetty Canal
Fig 20

November John Vickerman had been appointed Treasurer and John Disney had participated as a new shareholder.[21]

Rhys William Jones soon won the approval of the proprietors, and at the annual meeting held on 4 June, 1832, he was appointed "Principal Clerk, Clerk of the Works and Receiver of the Tolls", resident at Saundersfoot. On the same day he gave a report on the railway. "On the main line", he said "we have put in five culverts and formed the most difficult parts of the road through 'cutting' and 'banking' for the following distances: from the harbour at Mrs. Hughes' house to the foot of the inclined plane—820 yards; along the inclined plane and in the Crow Meadow above it —536 yards. In the Royal Park field beyond the incline 170 yards, in New Park field—240 yards, in Lower Cants field at the south end of the tunnel—280 yards. On Kingsmoor, on the north side of the tunnel—600 yards, near Begelly colliery—24 yards. Total = 2670 yards. The deepest cuttings ..... occur at Kingsmoor and at the south end of the tunnel, from which we have excavated over 10,000 cu. yards. The highest banking is on a culvert in the hollow (at Mr. Child's ground) between the Royal Park and the New Park fields which in one place is twenty feet high. On the branch towards Wiseman's Bridge we have formed the road from the harbour (at Mrs. Hughes' house) to the first or south point at Coppet Hall, 620 yards. A tunnel through this point 70 yards in length has been excavated, almost entirely through rocky ground, but in consequence of meeting with some old coal works in its course it has required arching for nearly half its length". The tunnel arch measured 8′ wide by 9′ high, which was thought to be ample for the traffic.

Jones went on to day that "that rails and blocks have been for some time in use at the breakwater and at other places where stuff from cuttings is carried out into the banks". They were permanently laid down at several places, including apparently the entire distance from Coppet Hall to the harbour. Blocks and rails were still being laid down "on the road from the harbour to the inclined plane" whilst on the Wiseman's Bridge branch "we are going on forming the cliff on the south side of the intended 100 yards tunnel through the second point at Coppet Hall and across the creek at that place". In conclusion Jones added that he was pressing on with the long tunnel at Kingsmoor.

A year later, at the annual meeting held on 3 June, 1833, the three principal shareholders Sir Richard Philipps, Charles Ranken and John Disney were all present. Edgar Disney also attended but evidently William Evans was no longer involved. As Sir Richard and Charles Ranken had subscribed for more shares the Treasurer was able to report that subscriptions totalling £12,000 had been received, and £10,623-1-6 had been spent. Construction work appears to have been progressing well, and it was agreed to adopt bye-laws for the use of the railway and harbour.

So far as can be judged the Saundersfoot Railway & Harbour Company began formal trading with effect from 1 March, 1834, notwithstanding the fact that by then the harbour had been in use for several months. By this date the railway was complete between Begelly and the harbour, and also between Coppet Hall and the harbour.[22] By 1 June, 1834, enough of the other outstanding work had been done to enable the company to seek a certificate of completion within the time allowed by the Act. At the annual meeting held on 2 June, Mr. Jones reported that the building of the line from Begelly to Thomas Chapel, and the remaining part of the Wiseman's Bridge branch had "for some time past been done, and we have this week finished laying down the rails on both lines ready for use". Apart from a few turnouts still to be made, and the relaying of portions of the track on stone blocks rather than wooden sleepers "from the nature of the ground", the main line and the branch had been completed. However the engineer thought that some fences should be finished, and gravel added to the horse path alongside the line. To do this, and to complete the works at the harbour, it was necessary to raise a further £2,500 in shares, thereby completing the sum of £17,500 allowed by the Act. This was done by the existing shareholders or partners. By the summer of 1834 they had good reason to be pleased: although some improvements were needed, the works authorised in 1829 were virtually finished, and trade was developing steadily.[23]

The Tramroad Era

The opening sections of the 1829 Act provided not only for the construction of the harbour and the Thomas Chapel main line, but also for two branch lines, one from Ridgeway near Saundersfoot to Merton (or Moreton) and the other from Saundersfoot to Wiseman's Bridge. As will be seen, the Merton (or Moreton) branch was never built. S.6 of the Act specified "that the distance between the inside edges of the said railway shall not be less than four feet, and the distance between the outside edges of the said railway shall not be more than five feet". The reluctance to stipulate an exact gauge seems to have contributed to a certain degree of doubt about the gauge of the line. The company's records usually referred to it as simply "four feet", but other reliable references mention 4′ 0⅜″ and 4′ 0¾″.[24] Bearing in mind the railway's somewhat unsophisticated track and rolling stock, it may well have been possible to find portions of line of all these measurements, but as the variations are modest 4′ 0″ no doubt sufficed in the company's papers. The gauge of 4′ 0⅜″ recorded by T. R. Perkins in the line's last years was probably the most accurate measurement .

Initially the railway was laid with cast iron fish-bellied rails 3′ 6″ long supported on stone blocks, some of which were cut roughly to 12″ × 12″ × 10″, and some of which were virtually uncut. In 1834 a section of timber sleepering on the Wiseman's Bridge branch had to be relaid with stone blocks because of "the nature of the ground". The original rails were supplied by the Yniscedwin Iron Co. and shipped from Swansea. After a number had been broken by cliff falls on the Wiseman's Bridge branch, 3′ 0″ long wrought iron bars measuring 3′ × 1′ in section were supplied by Messrs. Acraman of Bristol.[25] By 1840 there seems to have been a need to replace many cast-iron rails because of fractures in ordinary service. Even so, in recent times a number of fish-bellied and plateway ('L' section) rails have been found near the route of the line, and these probably date back to the very earliest days of the company.[26]

Although the 1829 Act allowed the use of locomotives, S.3 contained the interesting stipulation that "the furnace of every steam engine . . . shall be constructed upon the principle of consuming its own smoke''. No such locomotive was provided, and for many years the railway was worked by horses.[27] This was entirely appropriate because at the outset the line resembled a tramroad rather more than a railway. S.91 of the Act said more about the nature of the railway than any other. All persons might use the railway on payment of rates, it declared, ''provided the said waggons or other carriages shall not without Licence . . . pass upon the said railway at any other times than between the hours of Five in the morning and Eight in the evening during the months of November, December, January and February, and between the Hours of Four in the morning and Nine in the evening during March, April, September and October''. In the four summer months traffic was permitted between 4 a.m. and 10 p.m. The S. R. & H. Co., in common with many such concerns, provided the track but the users provided the ''waggons'' or ''carriages''. Although the company is thought to have owned some horses or mules, and suitable rolling stock, its primary purpose was to offer a transport facility to coal owners and others, rather than to run a regular transport service. Thus users placed their trams on the track, loaded them, and hauled them to the harbour, the horses walking alongside the track on a specially prepared path. So far as is known there was no timetable, and all traffic moved entirely as the individual users required.

In these circumstances the company was naturally interested in the vehicles brought onto the line, and their use. Under S.78 of the Act owners of waggons had to give a written account of lading, and S.81 spelt out a few basic requirements. It stated that the owner of every wagon should ''cause his, her or their name or names to be entered with the clerk of the company. Also such name, number and weight to be painted in large white capital letters and figures on a black ground, 2″ high at least and of a proportionate breadth, on some conspicuous part of the outside of every such wagon . . .'' The 1833 bye-laws underlined the requirements of the Act by providing that no wagon should exceed 3 tons weight when loaded, and that none should pass along the railway before it had been examined and approved by the company ''nor unless the wheels be of proper form, case hardened, and in the true gage of the road''. Little is known about the company's own rolling stock at this period, although there is a reference in the Minutes to the acquisition of ''stone carriages'' in 1837, and in February, 1840, ''train waggons'' were ordered for the transit of goods down the line. Another reference suggests that in 1849 the company owned no stock of its own.[28]

The 1829 Act and the subsequent bye-laws included provisions relating to tolls and charges, and the Act specified a fine of 40/- for any fare-dodging passenger.[29] However although the Act anticipated passenger traffic, the company never showed much interest —coal, limestone, ironstone, sand, timber and manure were the real business of the railway. Passengers may well have been carried in the first seven or eight years of the company's life, but an amendment to the bye-laws in 1841 put a considerable constraint on any idea of passenger travel. It provided that ''no driver or other person shall be allowed to ride on any tram, carriage or horse within the distance of fifty yards from any of the railway latches or shift rails.'' The bye-law added for good measure that ''no person under the age of 16 be permitted to act as driver on the railway''. This restriction became complete in 1857 when the bye-laws were further revised to declare that ''no driver or other person shall be allowed to ride on any tram, carriage or horse or other animal''. The same revised bye-law imposed a speed limit of 5 m.p.h. on the railway.[30]

Operating the line in the tramroad era seems to have been a slow business. Passing loops (quaintly called pass-byes) were provided at several places along the line, and the basic rule was that when two trains of trams met on a section of single track the lighter or unladen train would go back to the nearest passing loop. In the event of an accident or breakdown (and only in such circumstances) a driver was required to take the offending tram off the track as promptly as possible; penalties were imposed for the obstruction of traffic. Even so, the company had obstructions of its own in the form of a number of toll gates across the track to ensure that it obtained its dues according to the distance covered. Failure to shut any gate across the railway might incur a fine of 40/-. The incline inland from the harbour was another obstacle. The usual practice was for loaded trams descending by gravity to haul up empty trams. Until the incline chain was replaced in 1842 only two trams could be hauled in each direction at any one time, but thereafter the number was three. A path for horses and drivers was laid alongside the incline, but in wet weather this could cut up very badly.

The company's resources were devoted so completely to the construction of the harbour and the two railway routes that no attempt was made in the 1830's to build the branch line from Ridgeway to Merton, or Moreton. However, a very short branch seems to have been built to a quarry at what was then known as Betty Brown's point at Coppet Hall; it was recorded in June 1838, that this branch had been washed out by the sea.[31] In the same year an extension of the railway to other collieries at Begelly and Broadmoor was proposed; a short branch to the Begelly colliery from Barley Park (near Begelly church) was already in existence by this date apparently laid down by the colliery proprietors. In February 1840 the S. R. &. H. Co. found itself at odds with the Begelly company over the ownership of rails on this little branch at a time when the old cast iron rails needed replacement. Optimism was the mood of the moment, however, and in 1840 it was suggested that the railway should be extended to Ludchurch and even Prescelly Top. Nothing came of this notion.[32]

In spite of a good deal of earlier discussion, it was not until 1842 that the S. R. & H. Co. obtained further powers for extensions to their railway. An Act of Parliament passed on 31 May, 1842, gave the company formal powers to build (or re-build) the line from Wiseman's Bridge to the Lower Level colliery at Kilgetty, and renewed powers for a branch line from Saundersfoot to Merton—although in this instance the intended terminus was at Stonybridge, near to Moreton colliery.[33] The same Act also approved a branch railway running from a point on Kingsmoor Common near the Saundersfoot Railway's two milepost past Broadmoor colliery to Masterlands. In each case the company was required not to deviate from the authorised line by more than 100 yards, and the authorised levels by more than 5 feet. Special provision was made for the company to have an option to purchase any mines or mine workings within 40 yards of the railway route, but in the event of the company not exercising its option the existing owner was allowed to continue his mining activity.

In 1842 the railway was, on the face of it, poised for fresh expansion. In fact there was no further growth. No reference to the intended Broadmoor and Masterlands branch has been found later than June 1843,[34] and although the Moreton branch proposal was still being discussed in 1845 the very fact that the colliery was managing without it began to tell against the scheme. If Messrs. Williamsons, the proprietors of Moreton colliery, had been free to follow the route of their choice the line might have been built. The S. R. & H. Co., however, felt bound to insist on the alignment authorised by the Act. Accordingly the railway remained much as it was, although another new scheme soon appeared.

The Tenby, Saundersfoot and South Wales Railway

The years 1845 and 1846 were the years of the Railway Mania, the years railways caught the public imagination to such an extent that Parliament was overwhelmed with railway bills, some practical, some plausible and many improbable. Wales did not escape the excitement, and in June 1845, the Treasurer of the S. R. & H. Co. wrote enthusiastically about the merits of the proposed Welsh Midland Railway—a scheme to build a link from south and west Wales to the west Midlands via Brecon.[35] The S. R. &. H. Co. shareholders were less impressed, and preferred the plans of the South Wales Railway, authorised in 1845, to construct a line from Chepstow to Fishguard, via Swansea and Carmarthen, and including a branch to Pembroke Dock from the vicinity of Whitland. The prospect of having a main line railway passing within a few miles of Saundersfoot exercised a powerful appeal: minds immediately turned to the possibility of linking both Saundersfoot and Tenby to it by an additional line. A company was formed with the lengthy title of the Tenby, Saundersfoot & South Wales Railway and Pier Co. By October 1845 it was negotiating to buy the entire interest of the S. R. & H. Co. for £40,000, to enable the new undertaking to incorporate the Saundersfoot Railway into its route.[36] Agreement was reached on 6 August, 1846, the T. S. &

S. W. R.'s Act having received the Royal Assent on 27 July, 1846.[37]

So far so good, but the scheme had hardly got going before it ran into trouble. Famine in Ireland depressed hopes for profitable traffic through Fishguard and delayed the S. W. R.'s plans for railway building in Pembrokeshire. Eventually the Pembroke Dock branch was abandoned without any attempt to put a spade into the soil. The T. S. & S. W. R. was left like a bough lopped off a tree trunk. For a while the directors hoped for a change of fortune which might restore the tree and give their scheme some sustenance. They waited in vain, and although the S. R. & H. Co. relaxed the rate of interest on the purchase monies owing, by 1849 the T. S. & S. W. R. was hopelessly insolvent. The purchase of the S. R. & H. Co. was abandoned, and the plans of the T. S. & S. W. R. were quietly put on the shelf.

The failure of the T. S. & S. W. R. was not entirely bad news for the S. R. & H. Co. Mr. W. Chadwick and two or three of his friends amongst the promoters were also involved in the Pembrokeshire Iron & Coal Co.'s tremendous boardroom row. If the T. S. & S. W. R. project had succeeded the row would surely have spilt over into the affairs of the railway and harbour concern. The S. R. & H. Co. had worries as it was: a recession in the late 1840's created financial problems, and in spite of signs of improvement in 1850, matters did not get much better until the iron industry revived in 1852/1853. Even then the revival did not last, and the fortunes of the railway fluctuated according to the economic climate.

The 1850's and 1860's witnessed a gradual change in operations on the Saundersfoot Railway. If the distinctive feature of the tramroad era was the use of the line by independent operators using their own vehicles, the indications are that as collieries closed the number of independent operators declined. Early in the 1860's C. R. Vickerman became the dominant figure in the company, and by then there were very few others anxious to make use of the railway. Vickerman himself certainly felt free to consider the possibility of converting at least part of the line to the standard gauge, although the projects he preferred were soon defeated by the speed and resources of Messrs. Davies & Roberts, the contractors to the new Pembroke & Tenby Railway. Even so, the idea of a link between the Saundersfoot Railway and the rest of the railway system undoubtedly had its attractions. The 1864 Act of the P. & T. R. actually provided for a connection, but nothing was done to create it. Instead, to the chagrin of C. R. Vickerman, the S. R. & H. Co. had to make do with a siding for the exchange of traffic near the point where the P. & T. R. passed over the Saundersfoot Railway's tunnel onto Kingsmoor. In 1868, two years after the completion of the P. & T. R., this spot became the site for Saundersfoot station.

Decline in the local coal industry probably contributed to financial difficulties faced by the S. R. & H. Co. in the 1860's. On 5 August, 1867, the Court of Chancery appointed William Foley as Receiver of the company, C. R. Vickerman and George Williams acting as sureties.[38] On this basis trading continued until 1869, when the Court discharged the Receiver and put the company into the hands of Mr. Vickerman. Unfortunately for him, Edgar Disney, a relative of the former director John Disney, retained an interest in the company and evidently used it to challenge his authority. The matter went back to the Court of Chancery, and in February 1869, the argument was settled by an agreement that as holder of £6,200 worth of debenture stock out of a total of £8,000, C. R. Vickerman should purchase the remainder for £500 from Edgar Disney, who was the only other holder.[39] Thereafter the S. R. & H. Co. became an integral part of Vickerman's Saundersfoot empire.

The Locomotive Era

When C. R. Vickerman set in motion the events leading up to the formation of the B. C. C. & I. Co. in 1873, the S. R. & H. Co. was included in the scheme, apparently as a wholly owned subsidiary of the new undertaking. The setting up of the B. C. C. & I. Co. very soon brought the tramroad era to a close. At the period it was proposed that Kilgetty Ironworks should be given a new and direct rail link to the Pembroke & Tenby Railway at Kilgetty Station. To judge from a surviving plan, the branch was to run from a junction with the Saundersfoot Railway at a point just above the ironworks furnaces; it would then have curved and

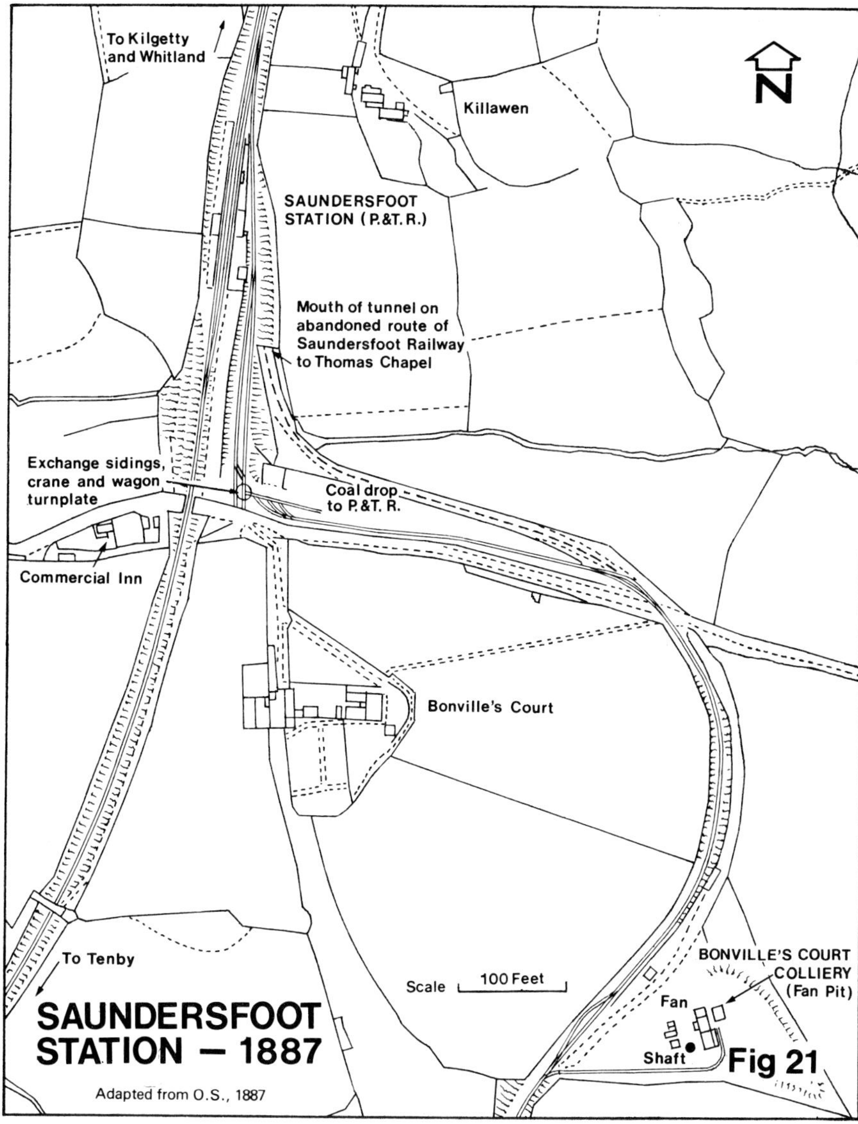
To Kilgetty
and Whitland
Killawen
N
SAUNDERSFOOT
STATION (P.&T.R.)
Mouth of tunnel on
abandoned route of
Saundersfoot Railway
to Thomas Chapel
Exchange sidings,
crane and wagon
turnplate
Coal drop
to P.&T.R.
Commercial Inn
Bonville's Court
To Tenby
Scale
100 Feet
BONVILLE'S COURT
COLLIERY
(Fan Pit)
Fan
Shaft
SAUNDERSFOOT
STATION — 1887
Adapted from O.S., 1887
Fig 21

climbed westwards to reach the village of Kilgetty then fast developing near the station.[40] However this scheme never materialised, presumably on grounds of cost, and the steps which were taken have the appearance of being an alternative. In October 1873, it was suggested that a steam locomotive should ben used on the Kilgetty branch of the Saundersfoot Railway, and as no real objection was made, the line between the harbour and the ironworks (and also the nearby collieries) was then relaid for locomotive use. In April 1874, the Manning Wardle O-4-OST ''Rosalind'' arrived on the railway.[41] Horses continued to work the short section between the harbour and the foot of the incline, and between Bonvilles Court and the exchange siding at Saundersfoot station. The decline of the collieries at Thomas Chapel and Begelly meant that little, if any, traffic was worked through the Kingsmoor tunnel.

The collapse of the B. C. C. & I. Co. caused further depression for the railway. In 1886 or 1887 the track west of Saundersfoot station was removed, and at about the same time more track was lifted in the vicinity of the now closed Kilgetty Ironworks.[42] A few years later, in 1893, the opening of the standard gauge siding into Bonvilles Court colliery caused the short portion of line from the Fan Pit to Saundersfoot station to be removed. On the positive side, the line between the harbour and the bottom of the incline was relaid for use by the ''Rosalind'' in 1889/1890.[43] By this date it is thought that there was little traffic from the harbour to the surviving collieries: the business consisted almost entirely of anthracite traffic from Bonvilles Court and Lower Level.

The end of mining at Lower Level colliery in 1900 appears to have been the occasion for the introduction of what was later and very loosely described as ''the Miners' express''. Following the colliery closure many men were offered alternative employment at Bonvilles Court, and to assist their transport they were provided with a special train consisting of the ''Rosalind'' and a number of empty coal drams. Between 6 a.m. and 7 a.m. each working day this travelled from Stepaside to Saundersfoot and then to the incline, at which point the men got out and completed their journey on foot. A return service operated in the evening between 4 p.m. and 5 p.m. The nickname of ''the Miners' express'' is usually attributed to a postcard publisher: the train was certainly the most photographed feature of the Saundersfoot Railway! Locally, however, the locomotive and train seem to have been known as "the Armour" on the basis of some resemblance to the armoured trains used by the British in the Boer War in South Africa. The miners themselves, though, were sometimes known as "the Boers"!

Towards the end of 1913 there was an encouraging development when businessmen interested in developing a new pit at Reynalton began discussions with the S. R. & H. Co. on the transport of coal. An agreement was sealed on 20 November, 1914, and the railway from Bonvilles Court to New Hayes, Thomas Chapel was rebuilt.[44] Heavier rails were laid on wooden sleepers, and the old stone blocks were buried beneath cinder and spoil waste ballast. An extension of about 1½ miles from New Hayes to Reynalton was also constructed, the route following the contours to minimize expense. To work the line the New Reynalton Anthracite Colliery Co. acquired the O-4-OST "Bulldog", and for safety's sake warning notices were erected where the line ran near or over public roads. As there was only one locomotive above the incline, and one below, signalling was still considered unnecessary. Even so, the S. R. & H. Co. was much improved by this period of modernisation. The line to Reynalton opened to traffic before the end of 1915.

The new undertaking soon ran into difficulties, and in May 1917, the S. R. & H. Co. noted that the company had made a "heavy annual loss". Rail traffic ceased in December 1917, and did not resume until after the end of World War I.[45] A further period of activity in the years 1919-1921 was scarcely more successful, and Reynalton colliery was abandoned in 1921. The N. R. A. C. Co. then proposed the sale of all the rails on the Thomas Chapel line, and extension, and the division of the proceeds between the two companies. The S. R. & H. Co. directors disagreed, and at a board meeting held on 5 June, 1922, made a counter proposal that the N. R. A. C. Co. should be free to remove 1,000 yards of track on their giving up all claims they might have on the remainder of the rails. This was accepted, and in 1922/23 1,000 yards of track at the Reynalton end were removed. The rest of the line from a point near Hacket remained in situ for several years, but was lifted west of Broom by 1936.[46]

After the closure of Reynalton colliery the locomotive ''Bulldog'' was transferred to the Bonvilles Court Coal Co. for use between the top of the incline and the exchange siding at Saundersfoot station. Even this restricted activity was stopped when Bonvilles Court closed in 1930, and for some while ''Bulldog'' and ''Rosalind'' languished in their respective locomotive sheds. In 1934/35, however, the revival of the coal industry brought both engines back into service. ''Bulldog'' operated between Bonvilles Court and Broom, and by 1936 ''Rosalind'' was at work on the line between the incline, the harbour and Kilgetty. Throughout their careers these locomotives were separated by the incline, and so far as is known they met only once—when the ''Bulldog'' was lowered down the incline and sent to ''the Platform'' at Setpaside for the removal of its cab roof.

The revival was regrettably brief, and when both collieries were abandoned early in 1939 the railway soon closed as well. Although it is thought that the railway carried some commercial traffic until May 1939, no record has been found of an official closure date. By August 1939, dismantling had begun, the first section to be lifted apparently being that from the incline to Saundersfoot station and Broom. Some of the track below the incline and on the route to Kilgetty was intact in 1940, and it was not until 1955 that the last sleepers were uprooted in Railway Street, Saundersfoot. By then Railway Street had been renamed ''The Strand'', and the route as far as the ironworks site (including the three tunnels at Coppet Hall of 70, 20 and 110 yards) had been converted into a public footpath. In spite of falls of rock from the cliffs and some heavy repair bills, the footpath has been maintained for the enjoyment of residents and visitors alike. Much of the Thomas Chapel route (including the incline) has also been declared a public footpath, although in 1964 a Birmingham based firm named Minitrains produced a scheme to relay the line as a miniature railway between Saundersfoot station and Begelly. This idea soon faded away, and now the tunnel below the station is blocked and the bridge across the main road at Begelly has been reduced to a solitary abutment.[47] After an active life of just over a century the Saundersfoot Railway has plainly passed on into history.

Locomotives and Rolling Stock

"Rosalind" An O-4-OST built by Manning Wardle of Leeds (Works No. 476 of 1874) for the Bonvilles Court Coal and Iron Co. Ltd. It had two outside cylinders measuring 8″ × 14″, and driving wheels of 2′ 6″ diameter. In its original condition it had a trapdoor front to the smokebox, and is believed to have been known by the name "Bonville".[48] C. R. Vickerman probably gave it the name "Rosalind" (after his daughter) soon after he had reacquired the company's assets in 1878. After the turn of the century the locomotive seems to have run without any nameplate at all. Throughout its career it was cabless, and fitted with a single central buffer at front and rear. A wooden locomotive shed was provided for it near 'the Platform' at Stepaside. The "Rosalind" scrapped at Saundersfoot in 1939/40.[49] The engine's last driver was the well-liked Mr. Billy Williams; the very first driver, Thomas Phillips, was charged with indecent assault in December 1874![50]

"Bulldog" An O-4-OST built by Kerr Stuart & Co. of Stoke-on-Trent (Works No. 2401 of 1915) for the New Reynalton Anthracite Coal Co. Ltd., and taken over by the Bonvilles Court Coal Co. in 1921. It had two outside cylinders measuring 9″ × 15″, and driving wheels of 2′ 6″ diameter. The wheelbase was 4″ 6″, the boiler pressure 160lbs./sq.in., and the weight in working order 12½ tons (empty 9½ tons). There was 18 cu.ft. of fuel space, and the tank had a capacity of 250 gallons.[51] Four sandboxes were fitted, and as built the engine had buffer beams with two buffers. These were soon changed for single central buffers appropriate for the rest of the rolling stock. "Bulldog" had a height of 6′ only, giving it 6″ clearance in the Kingsmoor tunnel following removal of the cab roof. It was provided with a small corrugated iron locomotive shed near the Fan pit at Bonvilles Court. On the closure of the railway "Bulldog" was sold to the Llanelly steelworks, leaving by way of the G. W. R.'s Bonvilles Court branch on 31 July, 1939. The locomotive was broken up at Llanelly in 1951.

The steam crane This was primarily a piece of harbour equipment, because it was positioned more or less permanently on the siding along the inner pier at Saundersfoot Harbour. It was used mainly for the discharging of ballast. The crane may have been installed as early as the 1870's, in the period of the B. C. C. & I. Co., but the first record of it is dated 1886.[52] The steam engine had one pair of 4½" cylinders with a 9" stroke. The steam crane and its rather distinctive roof may be seen in many old photographs of the harbour. It was scrapped in 1939/40.

The Director's Saloon This is believed to have been a coal tram rebuilt with a timber superstructure for the personal use of the Vickerman family and distinguished visitors. Unfortunately no photographs of the vehicle have come to light, and it is not known exactly when the carriage was built and scrapped. It does not appear to have been in service during the last decade of the railway. However a description of the vehicle in use has survived in a letter written by the late Commander E. Stuart-Jones. "What heaven it was" he wrote "to get into the S. R. & H. Co.'s director's carriage complete with guns, cartridges, dogs and . . . (be) drawn by the good MW loco Rosalind—shades of Hean Castle in the 60's and 70's". On this occasion the carriage was used to convey the shooting party to and from Pleasant Valley.[53]

4' gauge trams These were very varied in age and origin, a few possibly dating back to the tramroad era on the railway. It is known that in 1872 the Kilgetty collieries and ironworks had 47 wagons "for shipment of coal and culm", each carrying about 1¾ tons and being fitted with a single central buffer at each end. At the same date the ironworks had 34 trip trams, each conveying about 3 tons of iron mine. In addition it had the use of 4 wooden stone trucks, 12 wooden iron carriages, 8 large mine trams, 25 small mine trams, 12 furnace "dandies", 11 cinder wagons and 1 trolley.[54] Most of this stock was probably used in the ironworks only. In 1872 the Bonvilles Court collieries had 16 wagons, each carrying about 1¾ tons of coal or culm.[55] Additional wagons or trams were no doubt in existence at other collieries.

By 1900 the overall picture is clearer: in that year the railway had 75 "iron colliery loading trams". In 1913 the number in service was 71. After World War I it is likely that no new trams were built, and repairs were effected by cannibalising damaged or life-expired trams. Even in 1939 a large number of these trams were still in service, although very worn. At this date there were at least two timber bodied wagons on the Saundersfoot Railway. Tram wheels were of varying sizes, simply on account of wear. Some trams had inverted 'V' springs. All are believed to have been broken up at Bonvilles Court.[56]

Standard gauge wagons Ten 8-ton wagons were built for the Bonvilles Court collieries in or soon after 1870. Nos. 1-5 were built by the Midland Wagon Co. and Nos. 6-10 were built by the Metropolitan Wagon Co. All were paid for by quarterly instalments over seven year periods ending in 1877. Five more standard gauge wagons were provided before 1900, and it is believed that these were constructed by the Midland Wagon Co.[57]

C The Pembroke & Tenby Railway

By 1858 it was clear that the South Wales Railway was never likely to build its branch to Pembroke Dock. In reaction to this failure local interests went to Parliament for powers to build their own line. In 1859 the South Wales, Pembroke & Tenby Junction Railway was duly authorised to make a standard gauge (4′ 8½″) line between Pembroke Dock and Tenby. The title of the undertaking was soon simplified to "the Pembroke & Tenby Railway", and after a year of immense activity by the contractors, Messrs. Davies & Roberts, the line was opened on 30 July, 1863. The section between Pembroke Dock and Pembroke opened on 9 August, 1864.

The P. & T. R. always hoped to extend their line to meet the broad gauge (7′ 0¼″) S. W. R., or even better, one of the S. W. R.'s standard gauge competitors who were struggling to gain access to Carmarthen. Accordingly, in 1864, the company obtained powers for an extension from Tenby to Whitland Junction via Saundersfoot, Kilgetty, Templeton and Narberth. The P. & T. itself had rivals for the building of this piece of line, but the energy and determination of Davies & Roberts carried the company to success. Rails were laid through Saundersfoot and Kilgetty early in 1866, the contractors giving local residents their first sight of a steam locomotive in action.[58] After further rapid

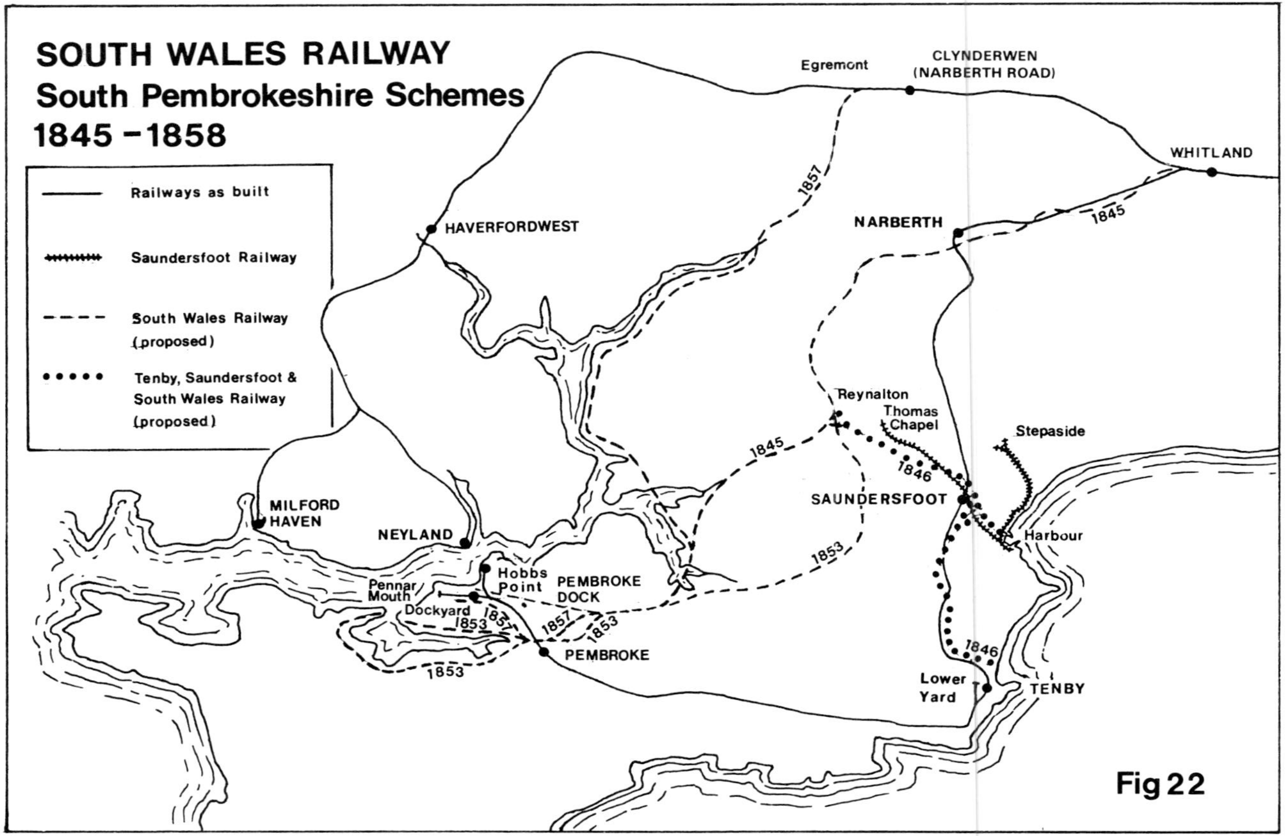

SOUTH WALES RAILWAY
South Pembrokeshire Schemes
1845 – 1858
Railways as built
Saundersfoot Railway
South Wales Railway (proposed)
Tenby, Saundersfoot & South Wales Railway (proposed)
Egremont
CLYNDERWEN (NARBERTH ROAD)
WHITLAND
NARBERTH
HAVERFORDWEST
1857
1845
Reynalton
Thomas Chapel
Stepaside
1846
SAUNDERSFOOT
Harbour
1853
MILFORD HAVEN
NEYLAND
Hobbs Point
PEMBROKE DOCK
Pennar Mouth
Dockyard
PEMBROKE
Lower Yard
TENBY

Fig 22

construction work, the Whitland extension was opened for traffic on 4 September, 1866. The company's hopes of going on to Carmarthen were eventually realised in 1868, by means of a track laid alongside the Great Western Railway (formerly the S. W. R.) between Whitland and Carmarthen.[59] The railway politics around this portion of line were complex, however, and soon after the Great Western abolished the broad gauge in South Wales in 1872, the P. & T. ceased to operate east of Whitland.

From Saundersfoot's point of view, the P. & T. R.'s most important Act was that obtained in 1864. It not only authorised the Whitland extension through the district, but provided for two short spur lines to connect with the Saundersfoot Railway.[60] In the event neither of these were built, probably because David Davies, the most prominent of the two contractors, had little regard for the local coal industry. At this very time he was investing heavily in new collieries in the Rhondda, and in comparison with that area of growth the Saundersfoot anthracite district was small-scale and showing signs of decline.[61] In addition it is possible that David Davies and C. R. Vickerman did not see eye to eye: although Mr. Vickerman gave supporting evidence before the House of Lords committees considering the P. & T.'s 1864 and 1866 bills, initially he had favoured other schemes.

When the P. & T. opened in 1866 a junction was provided for a siding to Moreton colliery, and Saundersfoot station was located nearby. Although Bonvilles Court colliery lay close to the line, it and other interests of C. R. Vickerman were ignored, much to his annoyance. After protests from him and other residents Saundersfoot station and its wooden hut were moved to the present more accessible site in 1868. Exchange sidings with the Saundersfoot Railway were installed nearby, but even then Mr. Vickerman was not entirely mollified. He wanted to see improvements at Saundersfoot station, and a siding laid right to the pithead at Bonvilles Court to avoid damage to coal when transferred at the exchange sidings. On both matters the P. & T. was reluctant to respond. Eventually in 1887 the company sanctioned the construction of corrugated iron buildings at both Saundersfoot and Kilgetty, and these were erected in the same year by the contractors Messrs. I. Dixon of Liverpool.[62] The Bonvilles

Court siding took longer to achieve, primarily because of argument over who should build the line, and who should own it. Agreement was reached at last in 1891, when it was decided that C. R. Vickerman and the Bonvilles Court Coal Co. should build the siding with no gradients exceeding 1 in 50, the P. & T. R. providing the junction and a stock of second hand rails.[63] The new siding was ready for traffic in April 1893, by which time the Moreton siding had disappeared. Following the closure of Moreton colliery it had been dismantled by December 1887.[64]

Up to 1870 the P. & T. was worked by the remarkable contractors, Messrs. Davies & Roberts. Although grateful, the company had to endure the strict Methodism of David Davies which prevented the running of Sunday trains. With the ending of the Carmarthen service in 1872 the P. & T. settled down to operating between Whitland and Pembroke Dock only. This restriction rather frustrated the hopes of the Barrow family, who bought their way into the company in the early 1870's apparently in the desire of fulfilling the old dream of making a railway from Manchester to Milford Haven. In this vision the P. & T. would have constituted the south-western extremity of the line; in reality it was already isolated by the Great Western Railway.[65] In time the inevitable happened, and the G. W. R. negotiated with the P. & T. for the take-over of the line. After leasing the P. & T. R. for a year, the Great Western absorbed the company completely on 1 July, 1897.

D Abortive Railway Projects

The Whitland and Milford Haven Railway 1862/63

This scheme appears to have been supported by several coal owners, including C. R. Vickerman, J. M. Child and T. Harcourt Powell. The Secretary of the promoters' committee was one George Clement.[66] Five new sections of standard gauge railway were envisaged, the first being a line from Whitland to Begelly, and the second a continuation from Begelly to Pembroke Dock. These portions were to measure 10 miles 3 furlongs 3 chains and 9 miles 2 furlongs 4 chains respectively. Railway No. 3 was intended to run from a junction at Begelly with railway No. 1 (which

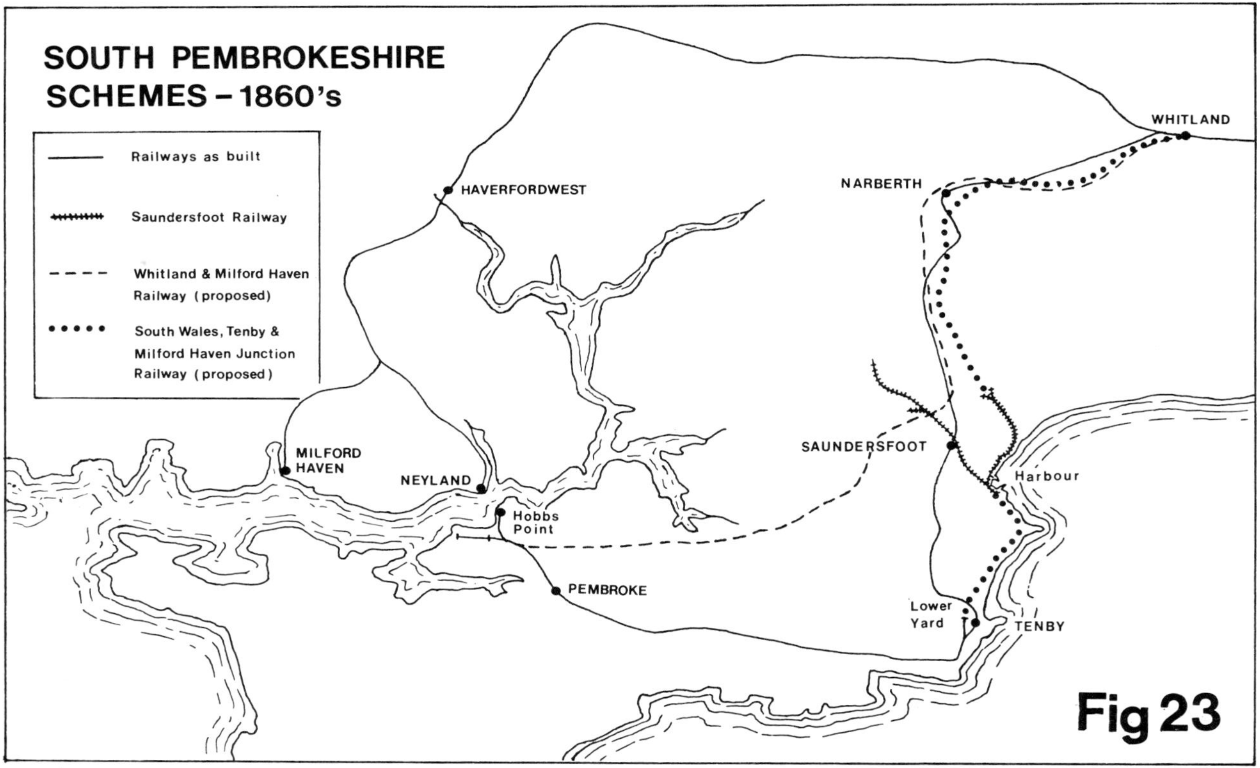
SOUTH PEMBROKESHIRE
SCHEMES – 1860's
Railways as built
Saundersfoot Railway
Whitland & Milford Haven Railway (proposed)
South Wales, Tenby & Milford Haven Junction Railway (proposed)
WHITLAND
NARBERTH
HAVERFORDWEST
SAUNDERSFOOT
Harbour
MILFORD HAVEN
NEYLAND
Hobbs Point
PEMBROKE
Lower Yard
TENBY

Fig 23

also connected with the Saundersfoot Railway) via the parishes of St. Issells and Gumfreston to a junction with the P. & T. R. at its terminus in Tenby. Railways Nos. 4 and 5 were shorter lines in the vicinity of Pembroke. Of the five lines only Nos. 1, 2 and 5 appear to have been surveyed in time for the Bill which was introduced into Parliament in November 1862. Even with these, it was said that some of the levels were wrong, and in February 1863, the Bill failed to pass Parliamentary Standing Orders. Opponents of the scheme (most notably the P. & T. R.) suggested that it would cost £400,000; James Shipway, the W. &. M. H. R.'s engineer, estimated that the works in the Bill would cost £240,000.[67] After the Bill failed there was talk of re-introducing it in the next session, and Messrs. Brassey & Field were mentioned as possible contractors.[68] There is no evidence that the scheme made any further progress, at least in this form.

The South Wales, Tenby and Milford Haven Junction Railway 1863/64

This project may have been the Whitland & Milford Haven Railway under an entirely new guise; it seems to have had the support of C. R. Vickerman and other coal owners. At any rate, a Bill for this line was prepared for the Parliamentary session of 1863-64, but never presented—probably because the progress of the P. & T. R. was so rapid as to make the Bill pointless. Even if it had been presented and passed it is debatable whether the capital would have been found and the railway constructed. It was an ambitious scheme, surveyed by the engineers C. B. Lane and Edward Bagot. Commencing at Whitland the line was intended to run via Narberth and Templeton before dropping down towards Stepaside. After passing near to the ironworks, the route closely followed the course of the Saundersfoot Railway. It swept around the back of Saundersfoot harbour and was then designed to hug the coast all the way to Tenby, tunnelling through the cliffs wherever necessary. Shades of the Great Western's sea wall at Dawlish! At Tenby it was proposed to reach the P. & T.'s terminus by tunnelling inland, the total distance of the line being 14 miles 1 furlong 6 chains.[69]

The Whitland, Cronware & Pendine Railway, 1877

This scheme was not of great importance to Saundersfoot because it was intended primarily to put Pendine and the Marros lime rocks onto the railway map. The route lay between the places named in the title, land over six of the seven miles being owned by one of the promoters, Morgan Jones.[70] When the project was first mooted C. R. Vickerman was induced to subscribe by an assurance that at a later date the line would be extended along the coast to Amroth and Saundersfoot. An Act to build the railway was passed in 1877, but by 1881 it was quite obvious that there was not enough money to build it. In 1882 the directors issued a prospectus advertising their revised intention to construct the line on the narrow gauge of 1′ 11½″, following the example of the Festiniog Railway in North Wales.[71] There was no further reference to Saundersfoot, and Vickerman withdrew. The directors failed to raise the capital to implement even the more economical proposals of 1882, and the scheme was formally abandoned in 1892. A subsequent proposal in 1893 for the "Carmarthen, Pendine and Pembroke Railway" made no progress.[72]

E Saundersfoot Harbour

Over the last 150 years the harbour has become both a symbol and a measure of Saundersfoot's fluctuating fortunes. When the coal and iron industries prospered, so did the harbour. When they declined the vessels became fewer in number. When the old industries vanished altogether, the harbour proved to be the last and most durable feature of Saundersfoot's past. To-day the harbour is as busy as the village, filled with visitors' boats and small fishing craft. It is the one aspect of industrial Saundersfoot with history still to make: it seems fitting to refer to it at the end of this study.

The formation of the S. R. & H. Co. has been described in Section B. According to the plans accommpanying the 1829 Act the original design for the harbour provided for only one long outer pier or breakwater. Construction was not far advanced, however, before it was realised that an inner pier was needed for greater protection. R. W. Jones' report on the state of the works

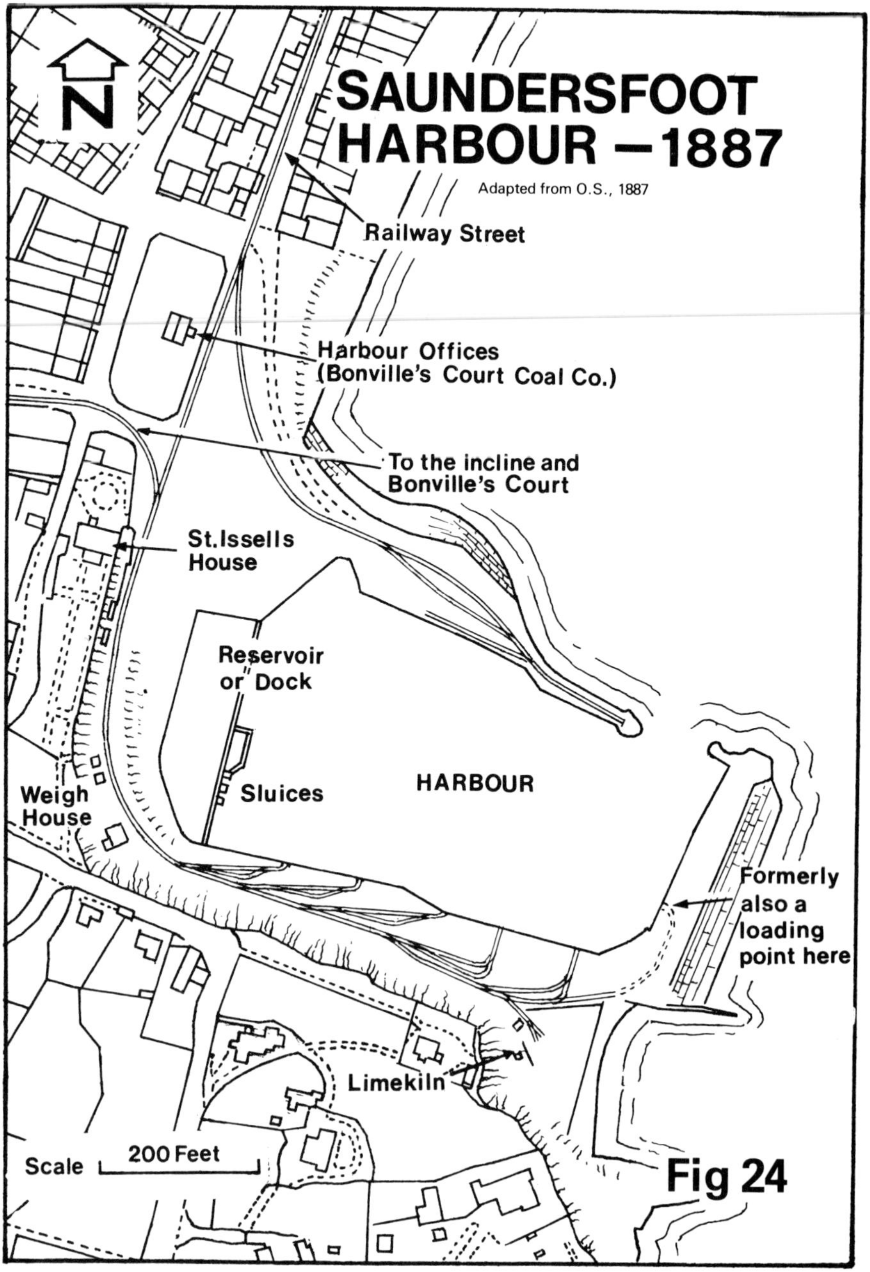
SAUNDERSFOOT
HARBOUR –1887
Adapted from O.S., 1887
N
Railway Street
Harbour Offices
(Bonville's Court Coal Co.)
To the incline and
Bonville's Court
St.Issells
House
Reservoir
or Dock
Sluices
HARBOUR
Weigh
House
Formerly
also a
loading
point here
Limekiln
Scale
200 Feet

Fig 24

in June 1832, refers to an intended northern pier, but most of his remarks dealt with the main breakwater. According to the report "The breakwater which was commenced at the extremity of the bank below the high cliff in December last is now carried out in its full height (7ft. above the highest tide) for a length of 65 yards and comprises stuff from the high cliff, from the harbour bottom and the massive stones procured at the Coppet Hall and Stammers quarries, and piles of beech timbers disposed to give the greatest strength." Jones added that he was pressing on with the breakwater in particular, and intended to terminate it "in a strong semi-circular head fifty feet in diameter formed of strong grouted masonry". As he had had difficulty in getting together sufficient teams of horses to carry stones to the breakwater, and had had to pay high prices for them, he recommended that the company set up a stable and buy some good horses for the work. He also recommended the purchase of a barge or lighter of about 25 tons to enable the works to be put into a safe state before winter. These proposals were approved.[73]

Progress continued, and in June 1833, the company adopted bye laws for the harbour and railway. They displayed a natural desire to keep a firm grip on the use of the harbour. Pilotmen were described as "hobblers" or "helpers", and the opening bye-law imposed a 5/- penalty upon anyone who acted as a "Hobbler or Helper of Ships" without being appointed by the company. The next bye-law required all hobblers to be recommended by five masters of vessels as to character and capability, and to be examined and approved of by the Harbour Master. Hobblers were meant to be men of merit! In emergencies each hobbler was allowed to employ an assistant, and a scale of charges for both was set out. From the outset it was specified that masters of vessels registering 30 tons and upwards entering the harbour should employ the first hobbler who offered—so creating a competitive attitude amongst the hobblers. Within the harbour, however, the hobblers' monopoly was complete, so slower men stood some chance of business![74]

Work on the breakwater went well, and by September 1833, the proprietors were discussing the provision of wharves and jetties at the harbour. In June 1834, Jones advised that "The harbour may be said to have been for some time completed sufficiently for the use of the vessels that frequent it. We have had two shipping jetties at work since the end of last summer (at which we are at present loading about two vessels per day) and we are now putting in a wall which we find necessary for the construction of a third''. In addition, the inner or north pier was being enlarged, more excavation was required to deepen the harbour, and the scouring basin was still to be completed. The scouring basin was a significant feature of the scheme, being a reservoir at the back of the harbour enabling silt to be removed by water action when the sluice gates were opened periodically. In spite of its importance, in 1834 some consideration was given to completing the basin as a dry dock. The notion was not pursued, probably because of the expense of installing dock gates instead of sluice gates.[75]

On 15th October, 1834, at Haverfordwest Quarter Sessions, the company claimed the certificate of completion of its works within the five year period specified in the 1829 Act. The harbour was then busy, but not finished in every detail. In May 1835, Jones noted that the scouring basin had only just been completed, and that a berth had to be made for ships discharging ballast at the north pier.[76] Unfortunately at the very time these improvements had been seen to, winter storms damaged the works. In February 1836, Jones wrote of damage at the shore end of the north pier, and of the problems with the harbour bottom. However he thought that repairs would soon be effected enabling them to go ahead with ''additional jetties and conveniences for trade''.[77] Repairs were carried out, but on 12th October, 1836, the harbour was hit by what Jones described as a ''hurricane''. ''Such was the fury of the storm'', he wrote ''that it dashed the big waves many feet above the whole breakwater, carrying the spray over the mast heads of the vessels in the harbour behind it . . . There were five or six vessels in the Harbour during the tempest including a sloop called ''the Wheatsheaf'' and a large brig of 400 tons burden called ''the Echo'' of Sunderland. The Wheatsheaf parted her moorings and was washed up behind the north pier, where she has become a total wreck. The Echo also parted some of her moorings, and has been damaged by thumping against the bottom and against the ballast birth where she lay.

The others fortunately rode out the storm without injury . . . The tide on that night as well as the wind was the highest ever remembered at Saundersfoot."[78]

Both piers suffered in the storm, but the proprietors agreed to the necessary repairs at a cost estimated by Jones at £600. They also agreed to spend £300 on deepening the harbour, and £200 on improvements including a harbour tide signal post and lamphorn at the head of the outer pier. It was resolved to let a piece of vacant ground behind the outer pier to the Begelly Co. for fourteen years from 1 January, 1837, at an annual rent of £15 per annum, the site to be known as "the Rock Wharf". Similar wharves were to be let to the Kilgetty Co. ("the Kilgetty Wharf") and to Morgan Hughes ("the Thomas Chapel Wharf"). Other parcels of land in front of the Milford Arms were to be let for warehouses to Mr. Sleeman of the Begelly Co. and Mr. Biddulph who acted (it is thought) for the Lower Level colliery.[79] Subsequently the harbour works were put in hand, and a small quarry was opened on the south side of Coppet Hall tunnel to provide more stone.

The year 1837 saw some changes of staff. With the completion of the works it was felt that the company required a competent clerk of works rather than a permanent engineer on the spot. Accordingly R. W. Jones' £60 per annum post was terminated, although the company declared it would be glad to have his assistance when required. James W. Crippen, who, like Jones, had been formally employed from September 1833, was also discharged, although on grounds of ill health. He had served as clerk to the company at Saundersfoot at £80 per annum, with the task of collecting tolls and duties, and completing monthly returns for the proprietors. In their place Samuel Giles was appointed "Collector and Clerk of the Works" with a salary of £150 p.a. His appointment took effect on 24 June, 1837.[80]

Most meetings of the S. R. & H. Co. took place at 4 South Square, Lincoln's Inn, London—the address of Ranken, Ford, Longbourne & Vickerman, Sir Richard Philipps' solicitors. However on 19 September, 1837 Sir Richard Philipps and John Disney met at the house of a Mr. Gower, in Tenby, and took the opportunity to review the financial state of the business. At the time the company consisted of three people—Sir Richard, who held 59 £100 shares, and Charles Ranken and John Disney, who held 58 £100 shares each. In addition to the total share capital of £17,500, the company had mortgage debts of £1,000 to Sir Richard, £1,500 to Charles Ranken and £500 to John Disney. This capital had been kept as low as possible as a matter of policy, and since the opening of the railway its receipts had been dedicated to completing the works. Treating 1st March, 1834, as the opening date, the gross proceeds in that year had been £516. In 1835 and 1836 the proceeds were £643 and £676 respectively, and in the first eight months of 1837 they totalled £654. Estimating the outlay over the period of £1,600 it was reckoned that £889 had been earned and expended on works. This represented a 3.6% return on capital. As trade was growing a dividend was hoped for in 1839. In the meantime John Disney appears to have worked hard to assist the company's prospects: in February 1838, the proprietors minuted their appreciation of his services. On the same ocasion the shareholders agreed to contribute an extra £100 each in anticipation of Parliamentary powers to create more capital.[81]

At the harbour the steady growth in trade was reflected by further improvements. In September 1837, it was resolved to enlarge the harbour's inner wall, to excavate the harbour bottom nearby, and to set up a weighing machine for trams. It was also decided to erect two new jetties with sloping and moveable shoots, and to alter the three existing jetties to the same design: these were to be numbered from 1 to 5. (Two more jetties were approved for the west side of the harbour seventeen months later).[82] By 1838 it was said that ships' masters spoke well of earlier improvements, including the night signal light. Sea swells still rolled into the harbour, however, causing injury to vessels, and an extension of the inner pier to narrow the harbour mouth seemed necessary. Unfortunately not all ships' masters proved to be responsible: on 11 November, 1837 Samuel Giles was concerned with the case of John Rees, master of the sloop "Peggy", who was convicted by Tenby magistrates of throwing ballast overboard into the harbour. He was fined 2/6, to the poor of St. Issells, and 4/6 costs.[83] Samuel Giles' other problems were problems of success: representatives of several collieries applied to build or to lease property on company land. In due course some leases were granted.

By 1839 the proprietors were hoping to obtain a new Act of Parliament. Apart from the need to enlarge the company's capital and to regularise the status of the railway already opened to Lower Level, the company wanted to develop the harbour as a "harbour of refuge", to make a wet dock (or floating harbour) capable of taking vessels of 300 tons burthen, and to take advantage of the extra tolls and dues payable. At the annual general meeting held on 3 June, 1839, it was noted that Thomas Dyson, Superintendent Engineer of the Bedford Level, was about to visit the district to direct new works at Kilgetty colliery. It was resolved to ask him to report on the best method of lenthening the inner pier, and to advise on the construction of a wet dock. Dyson acted promptly, and suggested that the harbour mouth be contracted by 110 feet, the inner pier being extended 90 feet. The cost he estimated at £400; the cost of the wet dock he put at £2,500. If properly done, Dyson declared, "it will arise Saundersfoot Harbour from the low station it now holds as one of the most unsafe, to one of the most safe harbours on the coast, and amply repay you for the outlay".

The shareholders of the S. R. & H. Co. agreed, and their ambitions were boosted by the annual report of John Longbourne, the Treasurer, on 1 June, 1840: "the success at this place has far exceeded the sanguine expectation which I held out to you in my report on 1 June last year, and the increase in shipping during the last five months and extent of preparations now being made by several colliery companies for working of their collieries afford very good prospects of the future shipment far exceeding those of any former year". He added that the extension of the inner pier would be completed shortly, and prejudice against the harbour was yielding. As regards the wet dock, local colliery proprietors had offered a guarantee of 7½% on the cost now estimated at £3,500, but as the scheme would require three years to complete Longbourne and Samuel Giles though a less expensive idea, like a floating gate for the harbour, would suffice. Francis Giles, the engineer appointed to deal with the works in the company's Parliamentary bill did not agree, and nothing was done.[84] In the meantime, whilst the possible contents of the bill were under discussion, the company decided to amend its bye-laws. The amendments reflected some of the practical problems that had arisen in running the harbour. Ships' masters now had to report particulars of their vessels on arrival, the Harbourmaster had to consent to the removal of any coal or culm from the quays or jetties, and "all idle or disorderly persons" were forbidden to trespass on or about the harbour.[85]

After some delay, the clauses in the draft bill were fully considered by the directors on 13 October, 1841 and 11 January, 1842. As the company was anxious not to be bound to complete the wet dock within a specified time, they decided to deal with the dock in a separate bill. The clauses relating to finance, mortgages, land purchase and the like were incorporated in the railway bill, whilst the fharbour bill provided for the wet dock and some extension to the company's powers in relation to the harbour. It put the cost of the new harbour works at £5,000.

The railway bill was passed on 31 May, 1842, and the harbour bill on 18 June, 1842.[86] In between, at the company's annual meeting on 6 June, the Treasurer felt obliged to advise the proprietors to postpone the new works for another year. Trade had dropped off in 1841, and John Longbourne now had to admit that business in 1842 was not up to expectations. Far from being optimistic about the future, he declared that he could see no prospect for improvement for several years. There was a glut of cheap coal on the market, for which he blamed Carmarthenshire rivals, and it seems freight rates were low, favouring suppliers of inferior coal. Keen to cut costs, the company discharged one man employed at the harbour, and reduced the wages of others.[87] A year later the Treasurer referred to the "horrible competition" which had developed between West Wales coal shippers, but acknowledged that the fall in the company's receipts had been trifling. The development of foreign trade was an important new feature: four cargoes of culm had been shipped to Alicante, and a sample cargo of about 40 tons was about to go to Prussia. In 1843 John Longbourne considered that if the collieries continued to prosper, and Mr. Stokes persevered with his plan to build a steam mill near Coppet Hall, it might be possible to establish a market in Saundersfoot and develop trade even further.[88]

The Treasurer's report in 1844 was similarly hopeful. Having mentioned a "very gratifying improvement" in trade, he observed that although local shippers were showing a "jealous avidity to sell" in competition with each other (to their own detriment) many old buyers had returned to Saundersfoot on finding the culm stronger than elsewhere. The Irish trade, in particular, was good.[89] In June 1845, business was even better, and the Treasurer then observed that the harbour was "full of vessels taking in coal, and two of the largest brigs are taking coal to Rochester". A great improvement in the iron trade was persuading Thomas Stokes to consider the best method of shipping ore from his estate, and this looked like a new source of income.[90]

Between 1845 and 1848 the affairs of the S. R. & H. Co., were overshadowed by the prospect of takeover by the T. S. & S. W. R. & P. Co. Although the scheme envisaged improvements to Saundersfoot harbour, remarkably little was said on the subject, presumably because the success or failure of the South Wales Railway was really the chief concern. In 1846, however, John Longbourne looked forward to the new project, saying "I entertain no doubt that ere long every spot of land in and near to Saundersfoot would be covered with houses if the Railroad and Wet Dock be constructed".[91] The signs are that the provisions of the 1842 Act for the improvement of Saundersfoot harbour were overtaken by the plans of the new company; the complete failure of the T. S. & S. W. R. & P. Co. seems to have put an end to the dream of a wet dock at Saundersfoot.

During this period Captain Thomas Stephenson was appointed Harbourmaster in place of Samuel Giles, who resigned in 1847. Stephenson also took on the task of making the annual report to the proprietors, and in June 1849, he drew attention to a gratifying increase in profits notwithstanding some extra outlay in repairing the coal jetties. He also mentioned the need to repair the pierhead and various items of equipment, and suggested an increase in the tolls on culm, and on harbour dues, in order to meet these expenses. The repairs were duly effected, and Stephenson's proposal initiated a revision in the company's bye laws, whereby the charge per ton register on ships entering the harbour was increased from 2d to 3d. A charge for the use of pier, light and signal was imposed on a sliding scale in relation to tonnage.[92]

As a result of the latest repair work the harbour was reported to be generally in a good state in 1850. There are no further references to the harbour for several years, but in 1857, it was noted that Thomas Stephenson had resigned and gone abroad, and John Morgan, his deputy, had been dismissed. Accordingly the company appointed William Foley as the new harbourmaster (he was already resident in the harbourmaster's house!) and John Child was appointed his deputy. Another revision of the bye laws was prepared, and these can be found in Appendix No. 4.[93]

Some measure of the development and use of the harbour can be gained from occasional newspaper reports on shipping at Saundersfoot. In May 1863, at a relatively prosperous period in the district, the Pembrokeshire Herald recorded arrivals and departures in one week, listing the name of each vessel, the owner, the port of origin or destination, and finally the cargo.[94] Shipping movements were as follows: **Arrived**—Sarah, Tucker, Bideford, earthenware; Heart of Oak, Jones, Cardigan, ballast; Industry, Williams, Porthcawl, wind-bound; Francis Tudor, Thomas, Carmarthen, empty. **Sailed**—Valiant, Morgan, New Ross, culm; Mary, Thomas, Port Talbot, iron ore; Thomas & Ann, Smart, Port Talbot, iron ore; Lady Kinsale, Gillard, Kingsbridge, coal; Mary Jane, Evans, Kilmore, culm; Eliza, Lewis, Dungannon, coal; Hope, Thomas, Cardigan, culm; Nice, Beeves, Fareham, coal; Sandwich Bay, Salisbury, coal and culm.

If the thirty years from 1845 to 1875 be regarded as the high tide of industrial Saundersfoot, it is worth noticing that the period corresponds with the heyday of Saundersfoot's very small shipbuilding industry. At some date in the early 1840's a shipwright named Francis Beddoe set up a boatyard in Railway Street, just above Saundersfoot beach and not more than 200 yards from the harbour. According to Barbara George, in the 1840's three vessels were built at Saundersfoot, together totalling a modest 83 tons. In the 1850's another three boats were built, totalling 104 tons, but in the 1860's only one ship was built, of 38 tons.[95]

During this period it is believed that Harry Louis Read opened his own boatyard in Railway Street, but is not clear to what extent he built new vessels or repaired older ones. All that can be said is that (according to Barbara George) in 1870's two more ships were constructed together totalling 107 tons. A photograph exists to show that one of these was built on stocks set up at the harbour. The last and largest of Saundersfoot-built ships was a cargo vessel actually named the "Francis Beddoe". It appears to have been laid down on the stocks in Beddoe's yard, but completed after Beddoe's death in 1873 by H. L. Read, the new owner of the yard. The "Francis Beddoe" plied local waters for many years before being wrecked on the treacherous Pendine Sands in 1920.[96] Thereafter the repair of coal schooners provided the mainstay of Read's boatyard business, although it seems certain that the yard also constructed some small sailing vessels and other tiny boats. Read's yard closed in or about 1910, but even then the skill of the shipwright did not completely disappear. Ever since at least a few people in the vicinity of the harbour have kept up the art of boat-repairing.

It was not easy in 1981 to imagine the hustle and bustle of the harbour over a century before. If shipbuilding seems improbable now, so does the rivalry of the "hobblers". Whatever the state of local industry, their competition to meet incoming ships seldom slackened. In his fascinating book on Saundersfoot T. G. Stickings quotes an incident in October 1876, when the pilotmen Samuel and David Darby vied with Edwin Phillips and George Childs to meet vessels approaching Saundersfoot in a gale. The account is well worth reading in full, but suffice to say that the Darbys having been disappointed in their efforts to reach two ships reached the third by a rather risky expeident. The vessels involved, two schooners and a sloop, were fairly typical of the craft making use of the harbour: they were the "Lewis" of Chepstow, the "James" of Milford and the "Sir William Molesworth" of Bideford.[97]

Some of the ships seen at Saundersfoot at the end of the nineteenth century have retained a place in local affections. Two seem to have been especially popular. The "Lady of the Isles" has been described as locally built, but her name was an indication of her origin—Jersey, in the Channel Islands, where she was constructed in 1868. Captained for many years by John Davies, this ship sailed regularly between Saundersfoot and Bristol, sometimes conveying passengers as well as cargo. In 1907 she suffered the indignity of running aground at Saundersfoot, but was refloated and sold not long after. She ended her days sailing out of Combe Martin, North Devon. The "Verbena" was another frequent visitor to Saundersfoot Harbour. A 93 ton ketch built at Milford in 1899, she was active for many years under the captaincy of W. J. Graham, who also owned the sailing ship "Advance".[98] Amongst the smaller craft, C. H. R. Vickerman's yacht "Ranee" was a common sight. The sporting interests of the Vickerman family extended to the sea, and they were instrumental in the setting up of the longshore sports at Saundersfoot Harbour late in the nineteenth century. This annual event, which was always conducted in a carnival atmosphere, died out with the advent of the Second World War.

In 1908 a dispute developed over the ownership of the foreshore at Saundersfoot. As early as September 1833, the proprietors of the S. R. & H. Co. had exercised powers in the 1829 Act to take possession of the beach between Monkstone Point and Wiseman's Bridge.[99] However when St. Issell's Parish Council asked the Board of Trade to grant a lease of the foreshore the Board claimed that the Crown had good title, and denied that the company (or anyone else) had any rights in it. The Crown even claimed foreshore rights within the walls of Saundersfoot Harbour, but as the Parish Council did not seek a lease of that area the claim was not pressed. If it had been, the issue would surely have led to litigation. As it was, the company appears to have conceded the leased area either to avoid local ill-will or for fear of the costs of a fight. The Parish Council then patted itself on the back, the Chairman (Alderman W. Lawrence) declaring in April 1909, that "coming generations would have something to be thankful for. If the Parish Council had done nothing else they had at least won the rights of the foreshore for the people of the neighbourhood".[100]

Between 1900 and 1930 the S. R. & H. Co.'s shares were held exclusively by members of the Vickerman family or their legal advisers, who continued to control the Bonvilles Court Coal Co. Ltd. Throughout the period the S. R. & H. Co. was heavily dependent on the B. C. C. Co. and its meetings usually took place at the offices of the latter company, near Saundersfoot harbour. World War I appears to have been an especially difficult time, and in May 1917, it was reported that the company had made a "heavy annual loss". Matters were no better two years later, and at the annual meeting held on 2 June, 1919, it was decided that as a result of a great increase in wages and materials, and a drop in traffic over the railway, the B. C. C. Co. should be notified that "it is imperative that they bear the cost of maintenance as well as the cost of structural repairs".[101] The directors present on this occasion were C. H. R. Vickerman (Chairman), his wife Eleanor G. Vickerman, and a relative Edmund J. Harvey.[102] The only other shareholders at the time were Frank Landon, the Vickermans' London solicitor, and his partner F. Murray.

Several changes occurred in the early 1930's. F. Murray and E. J. Harvey died in 1930 and 1931 respectively, and C. H. R. Vickerman died in 1932. He was succeeded as company chairman by his son J. F. ("Jack") Vickerman, who had joined the board in 1928. In spite of the closure of Bonvilles Court colliery, in September 1933, all 175 shared in the S. R. & H. Co. were vested in the Bonvilles Court Coal Co.[103] Although J. F. Vickerman continued as a director of this company Lord Merthyr appears to have acquired a dominant interest during the 1930's. This was a natural development as both he and Trevor T. Taylor (who was a fellow director) appear to have been better placed to invest in the revival of Saundersfoot's industry.[104]

The revival was brief, but for a few years the coal boats which had become very scarce after the Bonvilles Court closure, returned with greater frequency. A number of steamers were to be seen, including the "Combe" boats owned by the Ald Shipping Co. Ltd. of Bristol. The "Brockley Combe" (345 tons gross, 133 tons net) was a regular visitor, as was the "Castle Combe" (454 tons gross, 232 tons net) built at Bristol in 1936. They often took anthracite to St. Malo or St. Brieuc in France, or to Arklow or Galway in Ireland.[105] At the same period coastal sailing vessels still called to collect coal, and it is known that the ketches "Mary

Jane Lewis", "The Cornish Lass", "Annie", "Traly" and "Bessie Gould" were amongst those to use the harbour.[106] The S. S. Collin may well have been one of the last steamers to visit the harbour for anthracite: it is known to have left Saundersfoot for Bridgewater on 5 January, 1939.[107] There were other sailings at a later date, however, and at the time of writing the last reference found to a coal shipment from Saundersfoot relates to the M. V. Welcome which sailed for New Ross, Ireland on 11 May, 1939.[108] It is quite possible that local coastal craft sailed with small cargoes of anthracite after this date.

World War II saw the greatest changes at the harbour in a century. The railway was lifted, and the coal shoots were dismantled and sold for scrap, leaving the whole area quiet and desolate. All this was altered in the summer of 1943, when a vast military exercise was mounted along the coast from Saundersfoot to Pendine in preparation for the Normandy landings in 1944. The exercise, codenamed Operation Jantzen, involved 100,000 men including a large number of Americans.[109] It seems this did nothing to improve the harbour because by the end of the war parts of the structure required repair. Some work was carried out in 1946/47, but more money was needed. In 1949 Lord Merthyr sold all 175 S. R. & H. Co. shares to the Narberth Rural District Council for a nominal amount; early in the following year the local authority decided to try to seek a government grant to put the harbour into a good state of repair. In the meantime it was agreed to grant a seven year lease of part of the west wharf to the West Wales Fishermen's Association. By 1952 part of the harbour area had been converted into a car park, and the fees became the main revenue of the company. In June 1952, at a time when the harbour was under the direction of the St. Issells Parish Council, the route of the Saundersfoot Railway from Craig-y-Mor in the Strand to the bridge over the stream near Stepaside was conveyed to the Hean Castle Estate for the nominal consideration of one shilling. On 6 November, 1952, all the shares in the company were transferred to the estate company for the sum of 5/-, and in January, 1955, the shares were transferred again—back into the hands of Lord Merthyr.

This somewhat unsettled period of ownership was ended in

1958 with the issue of the Saundersfoot Harbour Order, which repealed the sections of the 1829 Act dealing with the harbour, and placed it in the care of the Saundersfoot Harbour Commissioners.[111] The Commissioners were given authority to employ a harbourmaster and staff, and all powers necessary for maintenance and administration of the harbour. Subsequently the Saundersfoot Sailing Club was allowed to build a new headquarters at the harbour near the site of the former weighbridge. Following the collapse of part of the inner wall of the harbour more repairs were carried out in 1963, and in 1969/70 a major programme of reconstruction was put in hand by the Commissioners.[112] In 1977 the Commissioners themselves had a new and rather striking harbour office building built at the north-west corner of the harbour, and this and the completion of re-surfacing work in the car park, meant that by the end of the 1970's the harbour was in better condition than for many years.

Footnotes

[1]For further reading on this subject see Barbara J. George's interesting booklet "Pembrokeshire Sea-Trading before 1900", reprinted from Field Studies Vol. 2, No. 1, (1964).

[2]PC 4082.

[3]T. O. 18.7.1867. Regular sailings between Bristol, Pembroke Dock and Ireland were advertised in the local press at the same period—e.g. H. M. H. T. July 1863.

[4]Information supplied by R. E. Bowen, who has made a special study of the industrial history of the Gwendraeth Valley.

[5]D. Williams "The Rebecca Riots", University of Wales Press, 1955. P. 181/2.

[6]At the time of writing an entirely new length of road from Killanow to Kingsmoor is proposed, passing to the north of Stepaside and Kilgetty. The narrow road through Kilgetty village, and restricted clearances on the old P. & T. railway bridge, appear to be the main reasons for the scheme.

[7]D. Williams: "The Rebecca Riots" p. 181/2. This excellent book is essential reading on this topic. "Efailwen to Whitland", Vol. 1, (1975), by the late E. T. Lewis, also contains some interesting material on the riots. According to a local story, construction of the new road was delayed for a full year by the need to gain possession of land occupied by the tips of a colliery close to what was once the Commercial Inn at Killanow. It is now known as the Stagecoach Inn, and overgrown tips can still be seen on the opposite side of the road nearby. Morgan maintains that the new road was first used in 1837. He also provides a most interesting description of the Begelly Turnpike Gate which was located alongside the Saundersfoot Railway's bridge over the new road.

[8]J. Copeland: "Roads and their Traffic, 1750-1850" David & Charles, 1968 p. 46/7. He quotes a reference published in "The Welchman", February 1843, to the effect that

each market cart had to pay the crippling sum of 12s. 6d. to travel 12 miles. Not surprisingly, it was asserted that the new road then saw very little traffic!

[9]Charles Hassall "General View of the Agricultural of the County of Pembroke", p.59, London (1794).

[10]Information supplied by R. E. Bowen.

[11]NCB/13540(C) shows the canal basin. P. H. Gosse: "Tenby, a seaside holiday" p.255, London (1856) refers to a ruined counting house near the shore at Wiseman's Bridge once used in connection with the coal trade.

[12]W. Davies: General View of the Agricultural and domestic economy of South Wales, London (1815) quoted by G. Edwards (1950).

[13]Goggerdan: Letter of J. Curr, 12.5.1797.

[14]Edwards (1950).

[15]PC 4098.

[16]The assertions were made to the writer by two elderly local residents in 1962, and cannot be checked. Some doubt has been expressed about the actual existence of this tramway, and although the evidence is not substantial the writer believes it to be sufficient. According to one local story the tramway was laid with wooden rails.; if true this would suggest that the line was laid in the eighteenth century. In any event the line wa almost certainly abandoned with the construction of the harbour and the opening of the Saundersfoot Railway.

[17]Hean Castle papers: extract from "Historical Account of Rivers, Canals and Railways of Great Brtain".

[18]10 George IV Cap. 108, 1829; House of Lords Committee Book, 22.5.1829.

[19]1829 Act, S. 40 and S. 41.

[20]Vickerman: S. R. & H. Co. Prospectus 5.5.1830.

[21]SRH 4.11.1831 and 11.1.1832. John Vickerman, father of C. R. Vickerman, was junior partner in the firm of Ranken, Ford, Longbourne and Vickerman. He died in 1834 as a result of being thrown from his carriage whilst on a visit to Sir Richard Philipps at Picton Castle.

[22]SRH 13.9.1833 and 27.1.1835. At the earlier meeting it was resolved to demand tolls and dues from 25.3.1834, but the date of 1.3.1834 mentioned at the later meeting is likely to be correct, being after the event.

[23]SRH 2.6.1834.

[24]T. R. Perkins: The Locomotive Magazine 15.9.1934; Railway Magazine, November 1936.

[25]SRH 5.6.1837.

[26]The plateway ('L' section) rails may have come from tramways underground.

[27]This stipulation was not unique—for instance, The Avon & Gloucestershire Railway Act, 1828, included a similar provision.

[28]SRH 5.6.1837, 7.2.1840, 4.6.1849.

[29]S. 133, 1829 Act.

[30]Bye-law revision approved by SRH, 27.1.1841 and 1.6.1857.

[31]This branch or siding was situated near the first tunnel at the end Railway Street (now The Strand), the junction facing the harbour. The quarry was very probably the one referred to as open in SRH 5.6.1837.

[32]SRH 1.6.1840. There was some point in proposing an extension to the limestone quarries at Ludchurch, three miles north of Kilgetty, but the supposed advantage of a line to the Prescellys is simply unknown.

[33]5 Vic. Cap. 35, 1842.

[34]PH 20.12.1863 reported a scheme to extend the line from Kingsmoor to Jeffreston and Loveston. This appears to have been a new proposal, and nothing came of it.

[35]SRH 2.6.1845.

[36]SRH 31.10.1845.

[37]SRH 10.8.1846.

[38]William Foley appears to have been clerk or secretary to the S. R. H. Co., having served with the P. I. & C. Co. for many years. George Williams role at this date is not known, although a gentleman of this name was clerk to the S. R. H. Co. in January, 1892.

[39]PRO, Chancery Lane, Ref. C33/1153. High Court (Chancery) 26.2.1869.: Vickerman v Saundersfoot Railway & Harbour Co., and Edgar Disney. Also T. O. 27.2.1869.

[40]Vickerman. Hean Castle estate plan, prepared by Messrs. Goode & Owen, Haverfordwest, c. 1872.

[41]T. O. 30.4.1874.

[42]Vickerman. Memorandum of 9.7.1886 shows track to be intact, but O. S. map of 1887 shows it to have been lifted.

[43]Vickerman. Hean Castle estate report and accounts, 16.7.1891.

[44]SRH 19.8.1914 and 20.11.1914.

[45]SRH 3.6.1918.

[46]SRH 5.6.1922, and 4.6.1923, T. R. Perkins—Railway Magazine, November 1936.

[47]The bridge over the main road at Begelly is believed to have been intact until about 1954. One of the abutments was removed during road improvements in 1978/79.

[48]Information supplied by E. S. Tonks and the Industrial Locomotive Society.

[49]It is believed that one builder's plate still exists in private ownership.

[50]T. O. 10.12.1874.

[51]Information supplied by E. S. Tonks and the Industrial Locomotive Society.

[52]Vickerman. Note dated 31.10.1886 records that Capital Hayward of the "Commerce" was knocked off the north pier by the tub of the steam crane, and fell between his vessel and the pier breaking two ribs.

[53]Cdr. E. Stuart-Jones letter to J. F. Vickerman (undated).

[54]Vickerman: Kilgetty collieries and ironworks, Schedule of moveable plant, 1872.

[55]Vickerman: Bonvilles Court collieries, Schedule of moveable plant, 1872.

[56]Railway Magazine, November/December 1946. Lower Level colliery is reputed to have had some timber bodied trams or wagons known as "Krugers", no doubt in contrast to the iron bodied trams or wagons usually incorporated in "the Armour", the train hauled by the "Rosalind". As in other parts of South Wales, colliery trams at Saundersfoot were frequently referred to as "drams". However company records invariably refer to "trams".

[57]Vickerman: Particulars of railway wagons, 22.4.1873. Also memorandum of 26.11.1900.

[58]The minute books and papers of the Pembroke & Tenby Railway are held by the P. R. O. at Kew.

[59]This was made possible by the conversion of the G. W. R.'s broad gauge up line to the 4′ 8½″ gauge, the first instance of broad gauge conversion in the country. When completed it provided two single tracks of different gauges on the same formation.

[60]Railways Nos. 4 and 5, P. & T. R. Act, 1864.

[61]The biography of David Davies is well worth reading: "Top Sawyer", by I. Thomas.

[62]P. T., 18.2.1887 records that the company chairman J. J. Barrow had received a petition from the inhabitants of Saundersfoot.

[63]P. T. 30.9.1891, 14.12.1891.

[64]P. T. 20.7.1887, 15.12.1887.

[65]An excellent account of the Manchester and Milford Haven dream is included in a recent book by J. S. Holden: "The Manchester & Milford Railway" Oakwood Press

(1979). As built the M. & M. R. comprised no more than 41 miles of single track in rural Cardiganshire, between Aberystwyth and Pencader Junction.

[66]H. M. H. T. 19.11.1862, 26.11.1862 and 18.3.1863.

[67]Records of the W. & M. H. R.'s brief Parliamentary career are held in the House of Lords Record Office.

[68]PH 22.5.1863.

[69]Some plans, and a list of landowners affected are held in the House of Lords Record Office.

[70]Bronwydd 6272/4/5.

[71]At this stage the company capital was stated to be £45,000 in 4,500 £10 shares, with powers to borrow £15,000 on the issue of debentures. The Festiniog Railway at this period was widely regarded as an advanced and well equipped narrow gauge undertaking.

[72]PT 28.2.1893.

[73]SRH 4.6.1832.

[74]SRH 3.6.1833.

[75]SRH 2.6.1834.

[76]SRH 1.6.1835.

[77]SRH 27.2.1836.

[78]SRH 4.11.1836. The letter from R. W. Jones dated 18.10.1836 was read to this meeting.

[79]SRH 6.6.1836.

[80]SRH 28.4.1837.

[81]SRH 19.2.1838.

[82]SRH 19.9.1837, 26.2.1839.

[83]SRH 19.2.1838.

[84]It is not known whether Francis Giles and Samuel Giles were related.

[85]SRH bye-law amendments approved 27.1.1841.

[86]The Act in respect of the harbour was 5 Vic. Cap. 53, 18.6.1842.

[87]SRH 6.6.1842.

[88]SRH 5.6.1843.

[89]SRH 3.6.1844.

[90]SRH 2.6.1845.

[91]SRH 1.6.1846.

[92]SRH 4.6.1849. Later in the same month it was reported that "a steamer from Bristol and Swansea would call at Saundersfoot if the owners could make the requisite arrangements". The proprietors agreed to let the vessel use the harbour "free of toll for present", but it is not known if any service was operated.

[93]SRH 1.6.1857.

[94]PH 22.5.1863.

[95]Barbara J. George: "Pembrokeshire Sea-Trading before 1900" Field Studies, Vol. 2, No. 1 (1964) p. 34 and 35.

[96]Howells, p. 7.

[97]Stickings p. 112-114, quoting The South Wales Daily News.

[98]Howells p. 7.

[99]SRH 13.9.1833.

[100]Narberth, Whitland and Clynderwen News, 22.4.1909.

[101]SRH 2.6.1919. Structural repairs following storm damage were not an uncommon requirement. A report and accounts dated 16.7.1891 indicate that considerable sums were spent on such matters in 1887-1890.

[102]Edmund J. Harvey was the son of Bridges Harvey, C. R. Vickerman's brother-in-law.

[103]SRH Transfer Register.

[104]From 1933 the board of the S. R. H. Co. (and apparently the Bonvilles Court Coal Co.) seems to have comprised the same three men who formed the board of the Kilgetty Anthracite Colliery Co. Ltd.

[105]Recollections of a resident at the period. According to W. T. 19.7.1934 ". . . on Saturday last the first cargo of 80 tons was shipped aboard the S. S. Staffa at the Saundersfoot quay. It is nearly 4 years since a vessel left the bay with a cargo of coal".

[106]27.4.1926, 6.5.1926.

[107]6.1.1939.

[108]11.5.1939.

[109]Howells p. 162/3.

[110]SRH 1.12.1949, 5.1.1950, 2.2.1950, 5.6.1952.

[111]Order made in pursuance of powers in the General Pier & Harbour Act, 1861, the Fishery Harbour Act, 1915, and the Sea Fishery Act, 1951.

[112]A plaque commemorating this reconstruction is located on the south side of the harbour.

Conclusion

When the naturalist P. H. Gosse visited Saundersfoot in 1854 he found a village dominated by the coal industry. "We passed through the little village of Saundersfoot", he wrote "with its small but strong harbour built of solid masonry. Several schooners were lying in it, and an air of bustle and trade prevailed, very different from the character of idle Tenby. All around in the country, tall scaffoldings mark the situations of the many collieries; and tramroads lead from the shafts to the harbour, by which the culm and the anthracite are brought down, and whence they are emptied into the vessels alongside the wharves through large wooden funnels". Gosse went on to mention the tunnels at Coppet Hall, the chimneys of Kilgetty, and near Hean Castle "a procession of carts loaded with wet coal-slack from the neighbouring mines."

The Saundersfoot Gosse saw has gone, and will not return. In many ways this is just as well, because for all the quality of the anthracite, the conditions endured to obtain it seem almost incredible now. After the Second World War the newly formed NCB made trial borings in the hope of finding fresh coal. The areas examined included Loveston, Reynalton, Thomas Chapel, Begelly, Cowsland (near Langdon), and Merrixton, but nothing was found in the right condition to justify further investment. In 1950 the coal industry was allowed to die, and in 1981 it seems safe to say that the only extensive area as yet unworked is in the coal measures beneath Saundersfoot Bay. Such are the costs and complications of coal mining in Pembrokeshire, it is hard to imagine that any attempt will be made to exploit this area for a very long time to come.

Although the coal industry in general clearly has a future, the most important source of energy in Pembrokeshire is now oil. Since the Second World War Milford Haven has demonstrated its potential as a deep water port, and has become a vital centre for oil refining and storage. This development has passed Saundersfoot by, and although tourism has made its mark on the village, the place surely shows more natural beauty than P. H. Gosse could see. That beauty is to be enjoyed to the full, but may industrial Saundersfoot, and all who served it, never be forgotten.

APPENDIX NO. 1

Issue of 16,500 Shares of £10 each,

of

THE BONVILLES COURT

COAL & IRON COMPANY LIMITED

Incorporated under the Companies' Acts 1862 and 1867, by which the liability of Shareholders is limited to the amount of their Shares.

SHARE CAPITAL £200,000, in 20,000 SHARES OF £10 EACH.

of which 16,500 are now for Subscription.

Price of Subscription, Par—or £10 per Share, Payable as follows:

£1 per Share	... ...	on Application
3 ,, ,,	... ...	on Allotment
3 ,, ,,	... ...	on 20th May, 1873
3 ,, ,,	... ...	on 20th June, 1873
£10		

Subscribers may pay up their Shares in full on any day on which an instalment is due, entitling them to discount at the rate of £6 per Cent. per Annum for such prepayment.

DIRECTORS

JAMES BEAUMONT, Esq., Merrow, near Guildford (late General Manager of The East India Iron Company).

WENTWORTH GORE, Esq., London, and Lyndhurst, Hants.

FREDERICK LEVICK, Esq., 46 Belsize Park, London (late Managing Director of The Blaina Iron Company)

'The Honorable' HENRY WILLIAM PETRE, Springfield, near Chelmsford, Essex.

CHARLES RANKEN VICKERMAN, Esq., Hean Castle, near Tenby, and Thoby Priory, Essex.

BANKERS

THE LONDON AND COUNTY BANK, 21 Lombard Street, E.C., London

SOLICITOR

JOHN HOLMES, Esq., 34 Clement's Lane, London, E.C.

AUDITOR

ALFRED ALLOTT, Esq., (Auditor of the Midland Railway Company)

SECRETARY

JOHN ROBERTS, Esq.

OFFICES

8 GREAT WINCHESTER STREET BUILDINGS, E.C., LONDON
BONVILLES COURT COLLIERIES, NEAR TENBY, SOUTH WALES

This Company is formed for the purpose of acquiring, working, and further developing the well-known extensive and valuable Coal and Mineral properties—portions of which have been in work from 40 to 50 years—known as the BONVILLES COURT COLLIERIES and the KILGETTY IRONWORKS AND COLLIERIES, situate at Saundersfoot, near Tenby, South Wales, together with the rich Hematite deposits at Penally, South Wales.

These extensive and valuable Works, upon which a very large sum of money has been expended, were formerly owned by Mr. C. R. VICKERMAN, in conjunction with several relatives and friends, but in consequence of deaths and family arrangements, the whole burden of the undertaking devloved upon him exclusively, and, as he resides for the greater part of the year in a distant county, he determined to dispose of the Works to the present Vendor to this Company, accepting, however, a seat at the Board.

The Company will therefore have the benefit of his valuable experience and assistance, which his knowledge of the district, and his influence with the resident population enable him effectually to give.

The Company also acquire the Railways (about 7 miles in length) which intersect the property, connecting it and the neighbouring collieries and works with the Harbour of Saundersfoot, and the Company also become Proprietors of the Harbour of Saundersfoot, which is of considerable local importance, being the port at which Coals are shipped from the district.

The BONVILLES COURT COLLIERIES are well known throughout England, as they supply the special quality of Coal chiefly used for malting purposes, called ''Myers' Malting Anthracite Coal'', which commands very high prices, being specially suitable for the preparation of the best malt.

The area of the Estates as to which this Company possess mining rights is more than 7,000 acres, or 11 square miles.

They are held upon mining leases for long unexpired terms, at very low dead rents and moderate royalties.

The Mineral lands comprise numerous veins of Argillaceous Ironstone of proved good quality.

The present contents of the Coal Fields are calculated at 8,000,000 tons. The Coal is of the rarest quality, is absolutely free from sulphur and other deleterious ingredients, and approaches more nearly to Charcoal than does any other fuel; and consequently commands very high prices at all times.

By the improved process adopted at the Kilgetty Works, the "Culm" is utilized, by conversion into a superior quality of Coke, for use in the furnaces; and it is specially suited for the manufacture of Patent Fuel, for which there is a great and continuous demand.

The deposits of Hematite Ore are rich in quality, yielding over sixty per cent. of Metallic Iron. The extent of these deposits not having been ascertained, no estimate has been made of their probable yield, although doubtless it will prove a source of considerable revenue. About 1,000 tons have been raised and used.

The Kilgetty Ironworks are of a very complete character, comprising blast furnaces, cast-house and hot-air stoves, gas apparatus, calcining kilns, coke ovens, engines and boilers, coal washing and grinding machinery, and all the requisites for the economical manufacture of Pig Iron, the existing furnace plant being capable of producing about 15,000 tons of Pig Iron per annum. It is intended to erect immediately additional furnaces, capable of treating a further 15,000 tons, thus increasing the total capacity of manufacture of Pig Iron to 30,000 tons annually.

The Pig Iron produced at Kilgetty is of special excellence, and is much esteemed from its similarity to Charcoal Iron, arising chiefly from the exceptionally pure character of the Coal used at the Works.

It is also specially suitable for the manufacture of Bessemer Steel, Armour Plates, Tin Plates, etc., and meets with a ready sale in anticipation of production.

The Works also include plant for the manufacture of "Patent Fuel".

Saundersfoot Harbour, with its quays, wharves, landing-stages, shoots, etc., was constructed under Special Act of Parliament, by which tolls and dues are authorised to be levied. These will belong to the Company, which will also have the right of levying a railway toll on all goods and traffic.

The possession of the Harbour, among other advantages, places the Works in direct communication with Milford Haven, the Government Dockyard at Pembroke, and by railway and sea with all parts of the kingdom. The saving of tolls and shipping dues alone averages from 1s. 8d. to 2s. 6d. per ton.

The Railways are about seven miles in length.

The purchase-money, payable by the Company for the entire properties, including the Railway, Harbour, etc. as a going concern, is £210,000, this amount being based on the Reports made by B. P. BIDDER, Esq., and LEWIS THOMAS LEWIS, Esq., F. G. S., jointly, and by Messrs. WILLIAM BIRD & CO.

As a proof of the confidence in the undertaking entertained by the Vendor, he has consented to allow the sum of £95,000 to stand as a deferred payment for five years, taking Mortgage Debentures bearing Interest at 6 per cent. per annum for such sum, the balance—namely, £115,000—being payable in cash.

This arrangement may be looked upon as of a very favourable character, as the entire profits of the undertaking—after deducting the amount necessary to pay interest upon the Debentures (amounting to £5,700 per annum only)—will be available for division as dividends on the Share Capital of the Company now for subscription.

After such payment to the Vendor, there will remain the sum of £50,000,—an amount considered amply sufficient to provide for the purchase of the stocks manufactured and now at the Works,—to furnish the necessary working Capital,—to provide for the development of the Coal properties,—and for such an augmentation of plant as to enable the Works to produce 30,000 tons of Pig Iron per annum, without interfering with the supplies necessary for the maintenance of the present Coal trade.

In the terms of purchase no addition has been made for the extremely valuable goodwill and brand, and the very valuable connection, extending throughout England and Ireland, both for Myers' Malting Coal, as well as the Kilgetty Iron, which is well known as of a specially choice make.

The current prices for the Coal delivered, f.o.b., in Saundersfoot Harbour is 27s. per ton. For Bessemer Pig Iron the price would be £8 10s. to £9 per ton; but as the present quotation may be considered, to a certain extent, as exceptional, the Directors have thought it safer to take, as the basis of their calculation, the estimates to be found in the following extract from the Report of Messrs. BIDDER & LEWIS, which, being based on an average of a number of years, may be considered normal:

> Dealing with an "output" of 200 tons of Coal and Culm per day = 60,000 tons per annum: at a cost of 8s. 6d. per ton; and assuming that one-half, or 30,000 tons, will be Culm [to be converted into "Anthracite Coke," and consumed in the blast furnaces, in addition to what is consumed by the engines, and sold to the local trade at cost price],

there will remain 30,000 tons, which, at the recent average selling price of 25s. per ton, would show a profit of 16s. 6d. per ton. But by allowing a further 2s. 6d., and not assuming any consequent increase in the selling price, there would still be left a profit of 14s. per ton	£21,000
The yield of four blast furnaces, at 150 tons each per week, would produce 30,000 tons, of Pig Iron per annum, which at present rates and prices, would show an actual profit of £2 to £3 per ton; but assume an average profit of 20s. per ton (which, for the special reasons already given, is fairly probable, and more likely to be increased than diminished), and the income from the four furnaces would give, per annum	£30,000
	£51,000

After providing for payment of Interest on the Debentures, the net profits would, according to these estimates, yield very large Dividends on the Share Capital of the Company.

An additional value is given to the position and extent of the property, and to the quality of the Minerals (before commented upon), by the very exceptional nature of the labour employed, as regards quality and permanence. It is drawn from the numerous class of small freeholders and occupiers of land, whose status forms a special feature of the district.

The Directors feel every confidence that the results of this Company will be highly satisfactory, the property being of a most complete character, offering a remarkable field for development in every branch of the undertaking.

Copies of the Reports and valuations of B. P. BIDDER, Esq., LEWIS THOMAS LEWIS, Esq., and Messrs. WILLIAM BIRD & CO., of the Agreement for purchase, and of the Memorandum and Articles of Association, can be seen at the Offices of the Solicitor.

In the event of no allotment being made, the deposit will be returned in full, without deduction.

Should a less number of Shares be allotted than are applied for, the deposit will be made available towards the sum payable on allotment.

Appliations for Shares must be made on the accompanying form, which must be fowarded, together with a deposit of £1 per share, to the Bankers of the Company, THE LONDON AND COUNTY BANK, 21 Lombard Street, E.C. London, or any of its Branches; or to the Secretary, at the Offices of the Company, 8, Great Winchester Street Buildings, E.C., London, where Prospectuses and Forms of Application may be obtained.

By order of the Board,
JOHN ROBERTS,
Secretary.

OFFICES OF THE COMPANY,
8 GREAT WINCHESTER STREET BUILDING, E.C.,
LONDON.
5th April, 1873.

The following Contract has been entered into on behalf of the Company:

5th April, 1873.—An Agreement made between THOMAS LEVICK of the one part, and JOHN ROBERTS, as Trustee for and on behalf of the Company, of the other part.

APPENDIX NO. 2

Memorandum dated 26th November, 1900 (attached to notes on Mineral Leases)

SCHEDULE A
BONVILLES COURT COAL COMPANY

Brief description of the Mineral District Collieries, Harbour and Railways etc. belonging to the Company

---

I Leaseholds

The property comprises the entire mineral District surrounding Saundersfoot Harbour producing Anthracite Coal or unequalled quality in addition to deposits of Argillaceous Iron Ore which yielded Iron scarcely inferior to Charcoal Iron and comprises about *4,000 acres* equal to *6 square miles* nearly.

The Collieries at present at work are two in number viz. *Bonvilles Court* and *The Grove* producing the noted Anthracite Coal called "Bonvilles Court" and "Kilgetty" commanding the highest prices in the Market.

Other Collieries not now at work but capable of being reopened at moderate outlay are also 4 in number viz. *Old Kilgetty, Foxholes, Moreton, Hacket* and *New Hayes.*

The Leases and conditions under which all the above Collieries are held are detailed in Schedule C. N.B. It is in contemplation to reopen the old Kilgetty Colliery which lies to the use of the Grove and Kilgetty Colliery and for this purpose to transfer the entire plant from the latter Colliery, which Colliery was closed a few months since. It was the coal from the Old Kilgetty pit which made the reputation of these Collieries 60 years ago and there remains a very large area unworked.

II Freeholds

1. The *Harbour* of Saundersfoot and *Railways* of about 7 miles connecting the collieries with the shipping Port, constructed and incorporated under Three Acts of Parliament viz. 10 Geo: C. 109 and 5 Vic c.c. 35 and 53 and entitled to charge the following *Tolls* and *Dues* viz. on coal and culm 2½d. p ton p mile Iron Ore 2½d. p ton p mile, corn shop goods etc. 8d. p ton p mile with 1d. p ton extra if conveyed in the Company's carriages.

Harbour Dues of 2d. p ton and shipping rates of 4d. p ton on Iron/10d. p ton and other dues as p Schedule to Act 5 Vic C. 53.

2. *One Third* share of *minerals* under *"Bickening Lands"* containing 18a. 2r. 30p. held on a tenancy in common with the Owner of the remaining 2/3rds about 4 acres of the Upper Vein have been worked.

III Capital Expenditure on Works

Since the formation of the Original "Bonvilles Court Coal & Iron Company Limited" in 1872-3 upwards of £44,000 have been expended in extensions and substantial and permanent improvements of the collieries, harbour, and Railways and Works, the details of which outlay are given in Schedule E.

Before any of this outlay hyas been incurred the works were valued by Messrs. Wm. Bird & Co. of Lawrence Pountney Hill, London on behalf and at the cost of Mr. James Carlton then an intending purchaser (who completed his purchase) at the sum of £225,000 as by their Report and Valuation dated 21st June, 1872.

IV Goodwill

The *"Kilgetty Vein"* has been known on the market for 200 years and the Bonville Court Vein for 50 years and upwards. Both veins have always commanded the highest prices 2% or 3% p ton above the *next* best qualities Messrs. Bass & Co. always take as much as we can spare them although they state that it costs them at their Maltings fully 5/- p ton more than "Gwaun-cae-gurwen" and other good Anthracites.

The Collieries have an old established connexion of very many years standing. The late Edward Chuck of Ware *then* the largest Maltster in England was one of their best customers, and put £10,000 into the collieries, on a change of ownership in order to "keep a hold" as he declared upon Kilgetty Coal.

The *"Bad debts"* during the past 7 years of depression have only averaged a loss between £9 and £10 per annum thus shewing the soundness of the business transactions.

V Completion

The gradual acquisition by the present owner as opportunities occurred during a period of 40 years and upwards of the entirety of the *Mineral District* and of the Harbour and Railways has given him the command of the Local Labour Shipping and Sales Markets and thus is avoided the disastrous competition which existed when there were 6 or 7 independent collieries in the District raising the cost of production and of shipment and lowering prices on sales. It also enables the owner to regulate the "output" . . . is of immense importance by diverting the "knocking down of prices" and the consequent difficulty of "getting them up again".

SCHEDULE D

SAUNDERSFOOT HARBOUR and RAILWAYS and Rolling Stock

Freeholds

Harbour with North and South piers of solid masonry of over three acres in extent with Ballast and Loading Quays 450 yards long.
Five (5) loading shoots with windlasses etc. etc.
There is sufficient depth of water for vessels of 500-600 tons burthen.
Reservoir (large) with double sluice gates divided from Harbour by substantial stone wall. The Company have powers under their Acts for converting it into a *Floating Dock* and charging extra rates for its use.
Steam Crane for discharging Ballast etc. with Engine of one pair of 4½ in. Cylinders and 9in. stroke.

Railways
1. From Grove and Kilgetty Collieries to the Harbour 2¾ miles in length—4ft. gauge laid with rails 42lbs. per yard.
2. Connecting Bonvilles Court and Foxholes Collieries with the Harbour and also with Saundersfoot station on the Pembroke and Tenby branch of the Great Western Railway—about 2,200 yards—same gauge.
3. From Hacket and New Hays Collieries to Saundersfoot Station about 2½ miles, at present unlaid with rails.
4. Self acting incline—with Drum brake and powerful chains between Bonvilles Court Colliery and the Harbour.
5. Rails and pass-byes around the Harbour and Quays.

Rates, Tolls and Dues chargeable under the Companys Acts of Parliament on

Coal and culm 2½d. per Ton per Mile
Iron Ore
Iron/10s. per Ton
Harbour Dues in registered Tonnage/2s. per Ton
Shipping Rates on Cargoes /4s. per Ton
Floating Dock (when constructed) 6s. per Ton

Rolling Stock Locomotive Engine 4ft Gauge 8″ Cylinders and 14″ Stroke built by Manning & Wardle, Leeds.
15 Eight Ton Railway Waggons—ordinary gauge built by the Midland Railway Waggon Company.
75 Iron Colliery Loading Trams—4ft. Gauge.

Bickening Minerals One third share of Minerals under A 18.2.30. (See Schedule A).

SCHEDULE E

Capital Expenditure from June 1872 to June 1878

	Grove Colliery (including pumps etc.)					959	10	10
b.	do Steam Capstan					642	14	0
c.	Bonvilles Court Colliery including alterations at Ventilator					740	8	10
	do (20″ Engine etc.)					567	17	8
	Railway and Harbour a/c including Locomotive and Steam Crane					8750	5	4
	Cottages					649	1	8
a.	Bonvilles Court Colliery Sinking shaft to win Kilgetty Vein					3318	9	3
	Stanley Minerals (freehold purchase)					325	0	0
						23743	19	4
	Expended by James Carlton under Agreements between 3,000 and 4,000 in 1872-73 (say)					3506	11	0
						27250	10	4
	Further Capital Expenditure from June 1878 to March 1889							
a.	Bonvilles Court Colliery completing sinking pit to K. vein	3017	6	4				
b.	Masonry Arching Roadway underground connecting Grove and Kilgetty collieries	1046	6	6				
	Fox Holes Colliery—Sinking Shaft 62 Fathoms	3328	3	5				
	Woodside Level Opening Work	334	6	11				
b.	Grove Colliery New Piston etc. to In Engine	267	7	6				
	Kilgetty Colliery—Nut coal Machinery	159	0	1				
	Bonvilles Court Colliery Pullies No. 2 & 3 etc.	1245	1	—		9308	1	9
	Add capital expenditure from March 1889 to September 1900					8009	0	0
	Total outlay—1872-1900					£44565	12	1

SCHEDULE F

COLLIERIES AND PLANT ABOVE AND UNDERGROUND

I Bonvilles Court Colliery

Landing Shaft 85 faths × 11ft. × 10ft. with pithead Weigh House capstan etc.

Winding Engine 24″ Diam. × 6ft. stroke in Stone built house with 2 flat wire ropes and 2 sheave wheels ft. Diam.

Pumping Engine 20″ Diam. × 6ft. stroke in stonebuilt house with 85 faths. of pump rods—29 faths. of 10″ pumps. 40 faths. of 11″ ditto 18 faths. of 5 ditto with poles polecases clack pieces etc.

Upcast Shaft 60 faths. × 9ft. × 10ft. with Waddle Ventilating Engine 20″ dia and Fan 16ft. diameter. Pit head pulley and guides complete.

Saw Mill and *Winding Engine* 16″ dia. with circular saw pullies wire ropes, bells, winding drum etc. complete in stone built houses.

Engine Houses Carpenters and Smiths shops offices stores and Lamp Houses stone built and slated.

Large Iron Water Tank and Reservoir

Revolving and Shake Screens at pits mouth for separating the coal from the nuts, peas and small culm.

Weighing Machines One farm weighing machine—one cart ditto—2 pit carts ditto.

Railway (4ft. gauge) to coal and culm works about 35 tons.

Boilers 2 5½ ft. × 20ft. galloway Tube Boilers
(9) 2 6ft. × 22ft. and 1 5ft. × 24ft. Tubular
1 4½ ft. × 22ft. - 1 3ft. × 24ft. 1 4ft. × 25ft. 1 4½ ft. × 33ft. cylindrical
All fitted with the necessary steam and feed pipes safety valves water gauges injectors boiler feed etc.

Hay and Chaff House containing Chaff cutting machine worked by vertical engine 8½″ × 1′ 8″ stroke.

Underground

Engine Slant Underground. 14″ Hauling Engine with about 1,200 yards of steel wire rope and sheave wheels and guides

One Tangyes' driven pump driven by a Priestmans Oil Engine

Pit and Drift to Kilgetty Vein Shaft 36 faths. and 11ft. dia. fitted with Boiler and Engine Room containing one new cylindrical Steel Boiler 4½ ft. × 30ft. complete with fittings.

One pair of 12″ *Winding Engines* Pit Head Steel wire ropes, 3 sinking Tubs, Water Tub etc.

64 yards of 3″ steam pipes and 72 yards of 4″ pumps in pit.

85 yards of 1½″ pipes 140 yards 2″ pumps in slope.

Slope Drift 114 yards long for winding and also giving second access to Kilgetty Vein.
Rails about 95 tons @ 16lbs. to the yard underground and on coal culm and rubbish tops.

Underground *Trams*		*166*
ditto	*Waterhaulers*	*6*
Screen Trams		*9*
Rubbish Top *Carriages*		*8*

1 *Horse Box* and 2 pit cages
Colliers Tools
Smiths Ditto
3 horses 1 Pony 1 Donkey underground with Harness etc.
Stables for 10 horses ditto

The Grove and Kilgetty Collieries

These Collieries are connected with each other underground by stone drift arched 870 yards in length the former taking the water of and ventilating the latter. N.B. It is proposed shortly to transfer the machinery from these pits to the Old Kilgetty Colliery.

II The Grove Colliery

Landing and Pumping shaft Bratticed in 3 divisions 105 faths. × 16½ft. × 12½ft. oval with pithead *40* feet.
Cornish pumping engine of 274 horse power 80″ dia. × 10ft. stroke in house, and 8ft. stroke in pit. (New piston and cylinder cover) working one 20″ pole one 22″ pole and one 12″ pole with 105 fathoms of pump rods 15″ square 35 faths. of 16″ pumps 51 faths. of 18″ pumps and 20 faths. of 12″ ditto.
Winding Engine 30 horse power vertical 40″ dia. × 3′6″ stroke fitted with Drum 15ft. dia. 4½″ wire ropes and 2 12ft. pullies.
Steam Capstan One new Loame's Capstan Engine. One pair of 12 engines and 120 faths. of 15″ Hemp Rope.
Boilers 4 double flued 7ft. × 30ft. in stone boiler house with slate roof
1 single ditto 5ft. 8″ × 30ft.
1 Cylindrical 4′ 6″ × 34′ 6″
all fitted with necessary safety and water gauges Steam and feed pipes etc.
2 Pit Cages each holding 2 colliery trams.
3 Engine and *Boiler House* Smiths and Carpenters Shops and Colliery stores substantially stone built and slated.
Incline (Self acting) fitting complete with Drum, incline ropes and *Carriages* each carriage holding

4 colliery trams 184 yards long—screens for separating coal and culm etc. etc. (114 yards double road and 70 yards single road)
Weighing Machines pit Bank and 1 Cart machine
Railway (4ft. gauge) to Coal Stocks and Colliery about 800 yards.
2 *Horses* and *Harness* underground
Stables for 10 horses ditto
2 Horses on surface with stables for same. Harness etc.

III Kilgetty Colliery

This is a modern colliery and is fitted up with nearly all modern Engines Boilers etc.
It is connected underground with the *Grove Pit* by an arched heading of 870 yards long.
Landing Shaft 84 faths. × 9ft. dia. with pithead and Guides complete.
Winding Engines 1 pair of 14″ × 3ft. in a substantial stone built house with 2 2½″ steel wire ropes 10ft. dia. sheave wheels and 2 pit cages 1 Air compressor for actuating underground pumps.
Slope Hauling Engine 16″ × 3ft. fitted with about 1,000 yards of 2½″ steel wire ropes 2 sheave wheels 7ft. diameter and 2 9ft. ditto in a stone built and slated house. One 12ft. 6″ × 24′ *Tangyes* Special *Pump* with 700 yards of 4″ pumps and 850 yards of 2½″ steam pipes.
2 *Weighing Machines* on Pit bank.
Screens for separating the coal and culm.
One revolving Screen Elevator and travelling belt for the manufacture of Nut coal and worked by a vertical 10″ × 8″ engine in house.
The whole covered in with corrugated iron roofing.
Engine Houses Smiths and Carpenters Shops Stores and offices.
Boilers 2 5ft. × 30ft. cylindrical
(4) 2 5ft. 8″ × 30ft. tubular
fitted complete with steam and feed pipes safety and water gauges etc.
Flooring Plates about 48 tons.
Storehouses and Workshops on Platform (by the Blast Furnaces) substantially built stone and slated houses comprising spacious offices, stores Carpenters Smiths and Fitters Shops containing
2 lathes (one very large and powerful) and 1 *saw mill* worked by *Engine* 11″ × 30″
Stroke and Boiler 4′ 6″ × 22′ 6″.
Fitters and Smiths Tools.
Cast House Spacious and lofty. Stone built and corrugated iron 52ft. × 69ft. and 27ft. high.

V Old Kilgetty Colliery

Engine Pit of 49 faths. × 10 feet diameter
Quarry Pit of 42 ,, × 8 —
Sunk to Old Kilgetty Vein of which a large tract to the deep commanded by a Bully driven from the Engine pit is still unworked.

VI Thomas Chapel Collieries

These Collieries command an area of over *1,000 Acres* and consist of the *New Hayes Pit* of 78 faths. × 10 feet diameter Sunk to the Kilgetty Vein of which but little has been worked and the *Hacket Colliery* of 29 faths. to the Lower Level Vein.
Coal of the finest quality has been worked to the rise from *Hacket Pit* but large and important mining can be made by driving a slope to the deep—or by a New Shaft at a spot indicated on the plan.

VI Foxholes Colliery

This pit of 60 faths. × 9 feet dia. is sunk to the *Timber Vein* also intersecting the *Rock* and *Low Veins.*

It is connected on the surface with Bonvilles Court Colliery by a slope tramway of about 350 yards.

One corrugated iron *Engine house* on stone foundation.

One 4ft. × 35ft. Cylindrical *Boiler* with steam feed pipes etc. complete.

VII Moreton Colliery

Two Pits to *Rock* and *Timber Veins* of about *50* fathoms each.

One pit ob about *150 fathoms* to *Lower Level* or *Bonvilles Court* Vein *Engine Houses* etc.

One 14″ × 24″ stroke *double winding Engine* with Drum and winding Gear nearly new.

APPENDIX NO. 3

February 1872

DESCRIPTIVE SCHEDULE OR INVENTORY OF IRON MAKING PLANT ETC. AT THE KILGETTY IRON WORKS PEMBROKESHIRE SOUTH WALES

Furnaces

TWO BLAST FURNACES in good repair and capable of being put into Blast at very short notice and at comparatively small outlay for Repairs.

FURNACE NO. 1 is 35 feet high, 12 feet boshes, Open Top fitted with Ring and Hopper and Gas pipes for taking the waste Gases to Engine Boilers and Hot Air Stoves and fitted at the Bottom with Blast pipes Valves and fittings complete and in working order—This Furnace is at the present time in good working order having been recently thoroughly and extensively repaired and requires but very little doing to before blowing in—when in blast she worked well and made the best quality and largest quantity in proportion to her size and the finest quality of Anthracite iron of any Furnace in Wales.

FURNACE NO. 2 is 53 feet high, 12 feet Boshes, Closed top, fitted with Cup and Cone and Rack and Lever for taking off the Waste Gases and fitted at the bottom with Blast pipes and valves. She is in good working order and having only been in blast a few months is as good as new, with the exception of the Hearth and Boshes which would need some little repair before blowing in.

Furnace Chimney or Stack

This is a very fine stack of large dimensions and was built for the use of Six furnaces consequently no further Stack will be needful if additional blast furnaces be erected.

Blast or Blowing Engines

There are 2 Blast Engines in the same house, Vertical Condensing fitted to work either together or separately.

No. 1 Engine has a 34 inch Steam Cylinder, Blowing Cylinder 64 inches diameter—and 7 feet Stroke.

No. 2 engine has a 30 inch Steam Cylinder Blowing cylinder 58 inches diameter and 7 feet Stroke.

Both these Engines are in complete working order. The Steam Cylinders are cased or jacketed to prevent the condensation of Steam and each Engine is fitted with a large and heavy fly wheel 16 feet diameter worked by a connecting rod from the Beam above.

Having been the original intention of the Proprietors to erect four Blast furnaces the Engine power was provided accordingly and is much more than sufficient for the present number of Furnaces.

In the same house is also a 24 inch Condensing Beam Engine 5 feet Stroke in good working Order with Fly wheel and Winding Drum for the purpose of taking such materials as may be required from the bottom level to the top of the Mine kilns and if not required for the above purpose this Engine is well adapted for working a Forge.

Boilers and Boiler House

There are Six large Boilers viz.

One 30 feet long	× 7	feet diameter	containing	2 Tubes
Two 30 feet long	× 6	feet ,,	,,	1 Tube each
One 30 feet long	× 6	feet ,,	Cylindrical	
Two 30 feet long	× 4½	feet ,,	,,	

They are all in good condition, and in working order they are fitted with large Steam and Gas pipes and adapted for being heated either by the waste Gases from the Furnaces or with small Coal or Slack. They are in one block, in a good and substantial house with Slated roof, and each Boiler being fitted with Steam, Safety, feed, and Water Valves can be worked independently of any other.

Blast and Water Mains

The Engines Furnaces and Hot Air Stoves are connected by a very large Blast Regulator 60 feet long by 7 feet diameter well stayed throughout and fitted with safety escape Valve and Patent Alarm or Telltale Whistle enclosed in a locked Box with weights to regulate to any desired pressure.

The Water pipes consist of two separate Mains leading from the Engines to the Reservoir; which is on a level with the Top of the Furnaces; in addition to a supply obtained from the neighbouring Collieries and a Main leading from the Reservoir to the Furnaces for supplying the different Tuyeres, Waterblocks, Boshes etc.

Hot Air Stoves

These are divided into four blocks viz. Two Blocks containing 63

Syphon pipes each, 7ft. high × 6″ × 3″ inside diameter divided into 3 rows of 21 pipes each. They are so adapted that the Blast can be heated either by means of the waste Gases from the Furnaces or by using small coal as fuel.

The two other blocks contain 12 Syphon pipes each 10ft. high × 12″ × 6″ inside diameter, divided into 2 rows of 6 pipes each; and are so arranged that the blast is heated by means of the waste Gases from the Furnaces.

All the Hot Air Stoves are on a level with the Tuyeres, and close to the bottom of the Furnaces.

Cast House

This embraces the frontage of the Two Furnaces and is very large and convenient and covered with a Semicircular corrugated Iron roof 67 feet long × 73 feet Span and 28 feet high supported by Cast Iron Pillars and Wrought Iron Stays. It contains a Casting Pit provided with a large and powerful Crane, Drying Stove, Two Cupolas, and sundry other Plant consisting of Flasks, Coreplates etc. and a very large and valuable quantity of Patterns, necessary for making large Foundry Castings. There is also a Reverberatory Furnace for warming and annealing Plates for making or repairing Boilers.

Refinery

This is built and fitted up upon the most approved principles with Cast Iron Moulds, Blast boxes, Water boshes, blocks etc. The whole being roofed over with corrugated iron supported by Cast Iron Pillars and Girders and capable of turning out 80 Tons of Refined Metal per week.

Mine and Lime Kilns

These consist of Two large oval kilns each 21 feet × 10ft. × 18½ft. high and Three Circular ones in good working order and situated close to the top of the Furnaces so that the charges can be drawn by the Fillers themselves and taken direct to the top of the Furnace. Situated in front of these kilns is a large corrugated Iron Shed supported by Cast Iron Pillars and Girders for depositing filling and weighing the Furnace Materials and is well placed for the mixing of the Furnace charges, if this mode of filling be adopted.

Smitheries

These for the use of the Furnaces and Foundry are very conveniently situated close to the back of the Furnaces and the Fires are blown by means of Pipes connected with the Blast Regulator.

Assay and Testing House

This contains a Crucible Furnace for Assaying Iron Ores—Testing furnace Slag etc. and can be driven either in connection with the powerful draught of the Furnace Chimney or by a pipe from the Blast Regulator.

Iron Ore and Limestone Depots

Large and commodious places for Stocking these materials are situated on a level with and close to the tops of the Furnaces and Mine kilns, in addition to which there is a large Depot where the Limestone is delivered by the Country Carts by way of back carriage for Culm.

Coke Ovens and Washing Machine

There are 20 Coke Ovens arranged in 3 blocks viz. One block of 12 Ovens, One ditto of 7 Ovens and a Single Oven with water laid on in front of each block. The whole capable of turning out *100* tons of Coke per week.

In connection with these Ovens there is a washing Machine together with Grinding and Mixing Rolls, for the purpose of washing the small Coal before Coking so as to keep the Coke as free as possible from all impurities worked by an independent 12 inch Horizontal Engine and Boiler all in good working Order.

A very valuable Furnace fuel which carried a heavy burden was made by the admixture of the small of the Anthracite with the slack of the Bituminous Coal. It was called Anthracite Coke and was used with very good and economical results in the Furnaces.

Fitting Shops, Lathes, Saw Mill Etc.

These consist of a large block of buildings divided off into Fitters Shop and Lathe room, Smithery, Carpenters' Shop and Pattern room.

The Fitters Shop and Lathe room contains one powerful self acting 11" Centre Lathe and One large and powerful Lathe, 4ft. centre for boring and turning large and heavy Castings, Outside and adjoining is a large self acting 4ft. Circular Saw Mill with Strapwheels and connections complete and in working order.

The Smithery contains 4 Smiths' fires blown by one of Lloyd's Patent Noiseless Fans, One Moulders Blacking Mill, Smiths and Fitters Benches and Vices, Tools etc.

The Carpenters Shop and Pattern Room contain One Carpenters'

Lathe and 4 Benches beside a very large and useful quantity of Patterns.

The above Machinery is all driven by an 11 inch High pressure Steam Engine with Boiler and connections complete.

There is also a large detached and convenient Saw Pit for hand or pit sawying covered in with Patent Asphalted Roofing felt.

Stores

Consist of 3 large, dry and convenient rooms suitable for storing the usual requisites necessary for the carrying on of Collieries and Iron works.

Offices

Consist of 3 rooms in one building for general and Private Offices and one detached for the uses of the Furnace Manager commanding a view of the works.

Manager's House

This is well and conveniently situated close to the works consisting of 17 rooms, Outhouses, Coachhouse and Stables and large and productive Garden with 3 to 4 acres of Pasture Land.

Furnace Manager's House

This is a substantial House built expressly for the Furnace Manager situated opposite the Works and commands a view of the whole.

Workmen's Cottages

These consist of one block of 6 substantial cottages built expressly for the use of the Men working about the Furnaces.

Moveable Plant

The Moveable and Rolling Plant consist of 4 large Stone Wagons, 8 large Mine Trams, 25 small ditto, 12 Furnace Trams, 11 Cinder Trams, 1 large Weighing Machine, 3 small ditto, 1 large scales and weights for weighing Pig iron, and about 30 Tons of Foundry Plant consisting of Foundry Ladles Flasks, Coreplates Corebars etc. etc.

Supply of Mine and Hematite Ores

The Argillaceous Iron Ore of this District is worked by means of Patching along a line of Cliff of about 2 miles and by Levels driven into the sides of the Hill of which the following are some of the principal viz. Cwms, Staggers Hill, Kilvelgy, Craigdam, Saundersfoot. When the

Furnaces were in Blast about 150 Tons per week were raised and by extending the workings and opening out others any requisite Increase could be obtained.

Hematite—a short time before the Furnaces were blown out a valuable deposit of Red Hematite Ore was discovered at Penally but it was only worked to a small extent and without the aid of Engine power. Analyses of the Ore is sent with this Schedule. The large Tract of Hematite Ground (about 2,000 acres) attached to these Iron Works is comparatively untried. The District is about 5 or 6 miles from the Furnaces and the Pembroke and Tenby Railway runs through it and gave a siding for the Transport of the Ore to the Furnaces. Other Deposits of Hematite Ore were discovered on adjoining properties at Lydstep and Jameston and Analyses of the Ores accompany this Schedule.

The Collieries and Coal Field

With their Engines, Boilers, Machinery and Plant are described so accurately and with such complete details in Mr. Blackwell's Report, and the Appendix thereto that it seems needless to repeat their description here. Since the date of that Report a large additional Capital has been expended on the Collieries and plant and in making further winnings by means of drifts to the South beyond what was contemplated when Mr. Blackwell reported upon them. These additional winnings have very much more than compensated for the moderate quantity of Coal which has been raised since the date of his Report.

Kilgetty Iron Works
Saundersfoot
South Wales

February 1872

APPENDIX NO. 4

SAUNDERSFOOT RAILWAY & HARBOUR CO.
BYE LAWS, AS REVISED 1857

BYE LAWS for regulating the Traffic and Management of the Railway

1. No Waggon or Carriages to pass along the line unless of a proper size and make suitable to the road and the wheels of the same to be of the proper form and size—case hardened—in the true gauge of the road
2. All Waggons or Carriages to be passed over the Inclined plane in such number order and manner as the Officers appointed to superintend the same shall direct and no Waggon or Carriage shall be passed over the Inclined plane unless in the presence of the said Officer to whom two hours notice shall be given to attend for that purpose
3. All Trams or Carriages passing along the Railway laden or empty shall be driven and brought on to the Weighing Machine for the purpose of being Weighed at the desire or upon the requisition of the Harbour Master or Collector
4. When Waggons or Carriages meet on the Railway the empty or lightest shall retreat or turn back to the next pass-bye or parting and each Driver shall place the rail latches in such position as will be required for his passing and so leave them and shall not drive along the Railway at a greater rate than five miles an hour, and no driver or other person or persons shall be allowed to ride on any Tram, Carriage, or horse or other animal
5. No driver shall leave his Tram or Trams upon the Railway and no incompetent person shall be permitted to drive on the Railway
6. If from any Cause a Tram break down or become incapable of passing along the Railway it shall be at once removed by the Driver to the side of the Railway so as to cause no obstruction to the Traffic and within twenty four hours afterwards the Driver or his Employer shall remove the Tram from off the property of the Railway Company
7. No Driver or other person shall haul any Tram or Trams off the Railway unless in case of accident
8. All persons not conforming to these Bye Laws or in any wise obstructing or molesting the Company's Officers in the execution of their duty shall be liable to all damages occasioned thereby and shall

be fined such sum as the Convicting Magistrate may think fit not exceeding £5 for each offence

9. All persons trespassing on the Railway will be prosecuted or otherwise dealt with as the Law directs And all Animals found straying thereon will be impounded

BYE LAWS for regulating Use of Management of the Harbour, Quays etc

1. All Masters of Vessels and all Hobblers, or Helpers of Ships acting as Pilots bringing Vessels into the Harbour with the Wind blowing W. to W.N.W. or outward from the Harbour, must come to an Anchor and warp the Vessel in, under the direction of the Harbour Master or his Deputy
2. All Masters of Vessels, Hobblers, or Helpers of Ships (in charge of the same for the time being) moving any Vessel in the Harbour, or taking any Vessel out of the Harbour under Canvas or otherwise shall be liable for whatever damage may be caused in consequence thereof to the Pier Walls, Jetties etc or to any Vessel, and be subject to a penalty not exceeding £5, in accordance with the Acts of Parliament relating to the Company
3. All Vessels within the Limits of the Harbour to be under the Control of the Harbour Master or his Deputy
4. All Masters of Vessels to stem and report the particulars of their Vessels upon their arrival within the Limits of the said Harbour
5. If any Vessel or Vessels cause damage to the Quays, Jetties or other Erections or to any of the Marks, Mooring Posts, Buoys or Works or to any other Vessel within the said Harbour the Owners and Masters thereof shall be liable for all such Damage whether through accident or otherwise in accordance with the Acts of Parliament relating to the said Company
6. All Booms, Yards, Spars and spare Anchors to be got in, peaked up and trimmed and all Ropes and Chains to be slacked at the Order of the Harbour Master or his Deputy
7. All Vessels to be in turn for loading or discharging in the order that they shall be reported at the Harbour Office on arrival and entered in the Stemming Book kept there for that purpose but no Shipper to occupy more than one Coal Jetty at a time, provided other Shippers are ready to load
8. Should the Vessel on turn not be ready the next Vessel on turn that is ready shall have the turn, the first Vessel on turn taking a berth next in succession if ready

9. All Ballast to be discharged and taken in at and from the places and in such manner as appointed by the Harbour Master and none other and by Men appointed by the Company at the rates named at foot of these Bye Laws
10. No Ballast to be thrown out within the limits of the Harbour except by permission of the Harbour Master
11. All Vessels with their Ballast or Cargo discharged and not going to load at the place of discharging, or not in turn for loading, and all laden Vessels, to move off from the Quays and Stages for others to discharge or load and all Vessels under the Jetties ready for taking in their Cargo or in course of taking in their Cargo shall move from one Jetty to another when required by the Harbour Master and in case of refusal or neglect the Harbour Master to send Men on Board to move the same at the expense of the Master or Owner
12. All Vessels waiting their turn for discharging, or loading to be moored and made fast at and to such places as the Harbour Master or his Deputy may direct and to move from one mooring to another when required by either of them to do so
13. Masters of Vessels shall keep sufficient hands on Board who shall hold themselves in readiness either to slack down their ropes or fastenings or move their Vessels as the Harbour Master or his Deputy shall direct
14. No Vessel shall lie waiting at a Jetty or Stage longer than one day Tide
15. On the rising of the tide, or as soon as any Vessel shall have completed her discharging or lading, the Consignee or Shipper or his or their Agent shall heave up the Shoots and properly secure the Machinery of the Jetty or Shipping place at which Vessel has been laden
16. Every Shipper or other person using a Jetty or Shipping place shall clear and remove from such Jetty or Shipping place or from the Platform of the same immediately after it shall have been used by him, all Coal, Culm Dirt, Rubbish or other Articles left thereon which shall be removed or deposited on obedience to the Orders and to the satisfaction of the Harbour Master
17. All Coal, Culm, Ore, or other goods or Merchandize of whatever description are to be placed and deposited on being landed according to the directions of the Harbour Master
18. All Coal, Culm, Ore or Goods of whatever description are to be removed by the Owners or persons in charge of them from off the Harbour Quays or Jetties within Forty eight hours after receipt of Notice from the Harbour Master to that effect And in default of

compliance with such notice Rent will be charged from the expiration of the Notice—but such Notice not to be given within 14 days from date of landing And if such Goods shall be in the way of other Goods they will be removed by the Harbour Master at the expense and risk of the Owner or Owners to another part of the Harbour

19. No Rubbish or filth shall be thrown into any part of the Harbour or the Basins, Ponds or Water courses belonging thereto
20. No person or persons shall pick up or take away any Coal, Culm, chips or other Articles from the Quays, Jetties or Harbour or any part thereof or the roads or approaches thereto without the permission of the Harbour Master
21. All idle and disorderly persons and all persons not having business to transact are forbidden to trespass upon the Quays, Jetties or Harbour or into the Storehouses or Yards or into the Office attached to the Weighing Machine or into any other Offices—or in any other way trespass upon the property of the Company and all such persons who shall be detected so trespassing may be taken into Custody and upon conviction before a Magistrate shall be liable to the fine mentioned in the Bye Law (No. 23) or otherwise dealt with as the Law directs And all Animals found straying on any part of the Company's property will be impounded
22. No person shall obstruct or molest the Harbour Master or his Deputy in the execution of his duty
23. All Masters of Vessels, Hobblers, or other persons not conforming to these Bye Laws shall be liable to all damages thereby occasioned and shall be fined such sum as the Convicting Magistrate may think fit, not exceeding £5 for each Offence

Harbour Regulations

Rate for discharging Ballast

1d per Ton for Wheeling (on the Register Tonnage of each Vessel)
For each Man employed on Board as may be agreed

Pier Light and Signal

per RegisterTons	to	30	Tons . . .	3d
30 ,,	to	50	,,	6d
50 ,,	to	70	,,	9d
70 ,,	to	100	,,	9d
100 ,,	to	150	,,	1/-
150and upwards				2/-

Rent: 1d per ton per week on all coal, culm, or other goods in accordance with Bye Law No.18.

APPENDIX NO. 5

NOTES ON THE COAL INDUSTRY AT LANDSHIPPING

For most of the nineteenth century the Saundersfoot and Kilgetty district could readily claim to be the most important coal mining area in Pembrokeshire. However in the earlier part of the century the Landshipping district offered some competition, although it was a smaller and more isolated district about twelve miles away to the north-west, on the banks of the River Cleddau. In its heyday Landshipping was in several respects similar to Saundersfoot: although more compact, the district was riddled with pits, tramways took coal to the shore, and much of the anthracite was exported by sea. At the beginning of the nineteenth century Landshipping was thriving, the coal industry making a contribution to the local economy possibly comparable to that of the Saundersfoot district. Even so, the story of coal mining at Landshipping has been curiously neglected; it is hoped that these far from comprehensive notes may do a little to enlarge interest in the subject.

Geologically Landshipping bears comparison with Saundersfoot, although no evidence has come to light to show that the deep Lower Level and Kilgetty veins were ever worked in this district. The existence of a fault in the coal measures (sometimes called the Old Mill Fault) confined most of the mining to the Landshipping valley between Landshipping Pill on the River Cleddau and Martletwy Wood, less than two miles inland. Little work took place south of this fault, and north of the fault the veins worked were, in descending order: the Rock vein, the Slate vein (sometimes known as the Tumbling vein), the Low vein (of the Timber vein group), the Big, High or Timber vein, the Little vein and the Stinkard vein. The Stinkard vein appears to have been worked in pits on the north side of Landshipping village, between Landshipping Ferry and Rigan Pill. The Little vein was never worked systematically, but the Big, High or Timber vein was found to correspond with the Timber vein of Bonvilles Court, and was the principal vein of the Landshipping district. The Rock vein was also worked extensively.

The coal industry at Landshipping is known to date from the seventeenth century, although there may have been some small scale mining in earlier periods. Before the end of the eighteenth century the devlopment of the industry seems to have brought about the construction of Landshipping bridge (across the pill) and also embankments running inland on each side of it. Landshipping quay probably dates from the same period. According to one local story, before all this building took place small lighters could come in on the tide to an inlet adjacent to

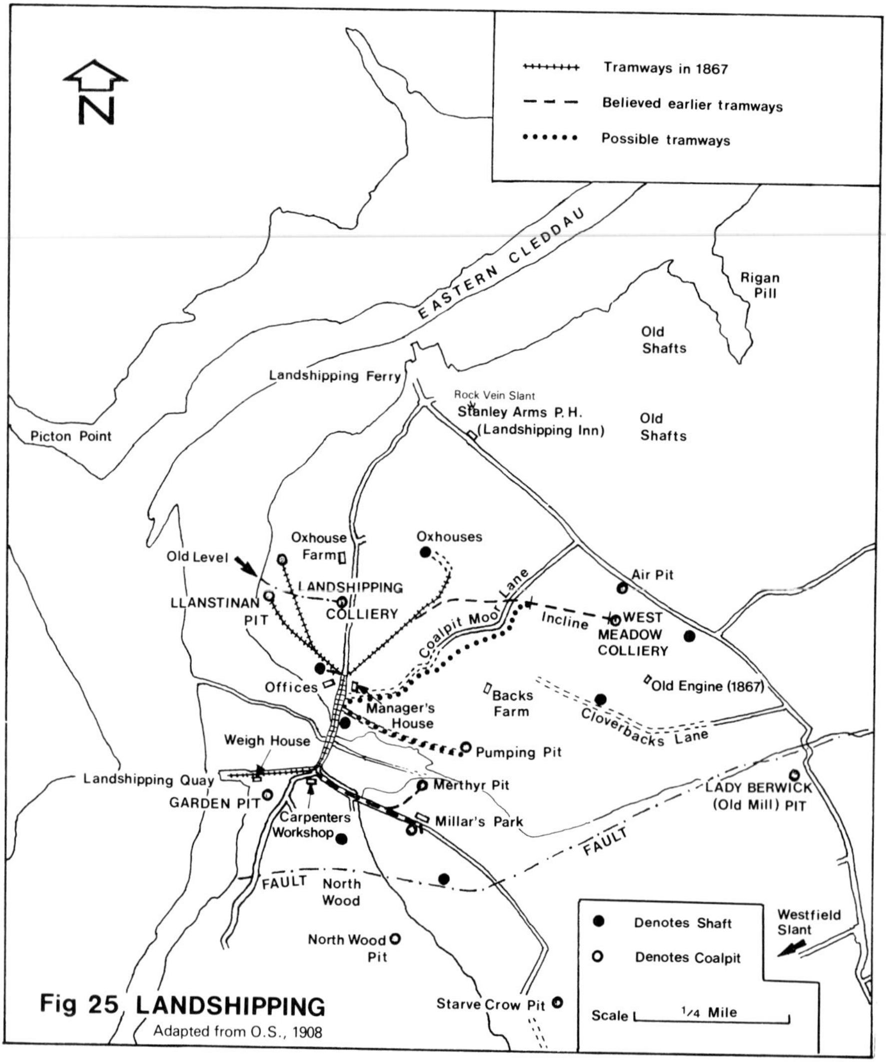

Fig 25 LANDSHIPPING

Adapted from O.S., 1908

North Wood in order to collect coal from nearby pits. As mentioned in Chapter 2, Sir Hugh Owen installed the first steam engine in the county at Landshipping colliery in 1800; unfortunately as the term "Landshipping colliery" appears to have related to different pits at different dates the exact site is not certain. Numerous maps show the word "Engine" close to the property called Clyne, between the Old Mill and Martletwy, but it must be doubtful if the original steam engine was situated so far inland. A map of 1867 shows an "old engine" on a site south-east of Westmeadow colliery; this may be a more likely location.

On a map dating from 1811-1820, the land on the north side of the Landshipping valley, from the bridge to the Old Mill, is covered by the words "coal works". This was clearly the principal area of activity at the time although the Garden Pit, on the south side of the Pill, was also in existence. Coal mining continued to flourish, and to judge from the report of the Children's Employment Commission, as late as 1842 Landshipping was the biggest colliery in Pembrokeshire in terms of manpower. At that time Sir John Owen, M.P., employed 163 people (132 males, 31 females) as compared with the 132 (86 males, 46 females) employed at the next largest colliery, Kilgetty. R. H. Franks received evidence on the colliery from Hugh Owen, and he noted that the pit employed three steam engines of 60, 16 and 4 horsepower. Access was by shafts 30 and 40 fathoms deep respectively (i.e. 60 and 80 yards), on which basis it might be surmised that the colliery described comprised the Garden Pit and its associated workings. The Garden Pit, close to Landshipping Quay, is known to have been 65 yards deep, whilst the Merthyr Pit, some three hundred yards to the south-east was 80 yards deep. The two shafts are said to have been linked underground, and it is quite possible that one or two other shafts in this vicinity may have been used as air shafts for these workings. All these workings were in the Timber Vein; this vein was reached at 72 yards in another pit, the Orielton Pit, 60 yards south of the Garden Pit.

The Garden Pit Disaster

The evidence to link the Garden Pit with the colliery Franks visited is by no means conclusive, but it is quite clear that the Garden Pit was active at this period. On 14 February, 1844, the pit gained enduring and unwelcome fame when water from the River Cleddau broke into the workings and drowned forty people. The Garden Pit disaster, the worst accident in the recorded history of the Pembrokeshire coal industry, has since become a subject surrounded by stories and legends.

The number of those drowned has provided one topic of controversy. The available records state that the number was forty, but figures ranging as high as seventy or eighty are sometimes quoted. The writer has even heard it suggested that the official figure was falsified because many of those who died were supposedly too young to be at work in the pit. No evidence has emerged to support such a contention. At the inquest on 27 February, 1844, it was noted that amongst the dead were fifteen boys aged 14 or under, only one of whom, a lad of 11, might be regarded as under age. There is no indication that women and girls were involved.

Another set of stories relate to the events preceding and following the disaster. According to one version, on the day before the accident a collier who was dumb spotted a small leak in the workings beneath the river and gesticulated to the overman that the mine was not safe. No action was taken, and so he tried in vain to pass on his message by scratching with a stone on the timber supports. Next day he was drowned. A much more probable account was given to Roscoe Howells by the late Bertie Scourfield, a survivor of the Loveston disaster, who had worked as a boy with old colliers who talked much of the Garden Pit accident. Apparently on the evening before the disaster the men were working so close to the bed of the river above them that they could hear the oars of lighters going downstream. They knew that it would be dangerous to go in the following day when there was going to be a very high tide, but they were virtually driven in by an unrelenting management.

The harshness of the management is a point generally agreed in these stories. It is said that after the catastrophe a notice appeared on the headgear of the Garden Pit saying, in effect, that the coal industry would never prosper. The management swiftly removed it, but the next night it appeared once again. Elizabeth Arthur, in her delightful memoir of Landshipping, "Turning of the Leaves", declares that the message was written in Latin and could only be translated by the parson. He found it to say that the coal industry would not prosper again "while this generation lasted at Landshipping". As Miss Arthur observes, there have been attempts at coal mining since but all have proved unsuccessful.

The Decline of the Coal Industry

Although the Garden Pit disaster administered a profound shock to Landshipping and district, it did not kill off the coal industry. Mining continued at other sites for some years until the Landshipping Colliery on the Llanstinian estate, north of Landshipping bridge, was put up for auction in 1857. It is thought that at the time the Owen family was deeply in debt. In any event, papers prepared for the sale stated that anthracite lay under nearly the whole estate in the parish of Martletwy, but only 50 acres out of 1,340 had been exhausted by the existing colliery over a period of seventy years. The existing workings were said to be very limited for want of capital, but with the "judicious application of